FINDING
OUR
TONGUES

FINDING OUR TONGUES

MOTHERS, INFANTS AND THE ORIGINS OF LANGUAGE

DEAN FALK

A Member of the Perseus Books Group
New York

Published by Basic Books
A Member of the Perseus Books Group

Books published by Basic Books are available at special discounts for bulk purchases in the United States by corporations, institutions, and other organizations. For more information, please contact the Special Markets Department at the Perseus Books Group, 2300 Chestnut Street, Suite 200, Philadelphia, PA 19103, or call (800) 810-4145, ext. 5000, or e-mail special.markets@perseusbooks.com.

Designed by Pauline Brown

Library of Congress Cataloging-in-Publication Data

Falk, Dean.
   Finding our tongues : mothers, infants, and the origins of language / Dean Falk.
      p. cm.
   Includes bibliographical references and index.
   ISBN-13: 978-0-465-00219-1 (alk. paper)
   1. Biolinguistics. 2. Language and languages—Origin. 3. Evolution.
I. Title.

P132.F35 2009
401—dc22

                                                        2008044299

10 9 8 7 6 5 4 3 2 1

For my granddaughter,
Eve Penelope Schofield—
the reader in our family

CONTENTS

PREFACE

How LANGUAGE BEGAN IS a deeply challenging question—intellectually, philosophically, and emotionally. Obviously, there is something about the subject of human origins, intelligence, and our uniqueness as a species that arouses strong opinions, and this can be seen in the polarized questions asked today: Did language begin millions of years ago and evolve slowly, or did it originate much more recently and abruptly? Was the first language (protolanguage) derived from the animal calls of our earliest ancestors, or from gestures? Did language evolve primarily for thought, or was it always used for social communication? And what is the relationship between language and the development of music?

Conflicting theories abound. I believe, however, that most researchers have been looking for the beginnings of language at the wrong point in time—they focus on the recent evolution of *Homo sapiens*, in the last 200,000 years—and thus have missed a crucial part of the puzzle. This book will take a longer view. The key to understanding the origins of language does not lie *after* the development of protolanguage, some two million years ago, but *before*—in the mysterious transition between our early ancestors' divergence from other primates, five million to seven million years ago, and protolanguage's first appearance.

More important, many researchers have missed clues about the emergence of language that appear every day in modern infants and toddlers. While infants are learning to speak, parents talk to them in a special way—known as baby talk, musical speech, or *motherese*. Some linguists have argued that motherese is not universal, but I'll show that motherese exists in *all* societies, sometimes in disguised forms tailored to various cultural taboos and practices.

One reason we've misunderstood the role of motherese in the development of language may have to do with assumptions about gender. Since at least Charles Darwin's time, men have been viewed as prime evolutionary movers because of their hypothetical focus on hunting, tool production, and warfare. More recently, women have also become celebrated as evolutionary movers because of their roles in gathering food and helping daughters raise offspring. Still, despite ongoing intense interest in language origins, there has been little focus on how motherese first emerged. This book shows exactly how motherese continues to help babies all over the world learn language, from the moment they are born until they are walking, talking toddlers—and what such observations can tell us about the emergence of language among our species.

From the fossil record of our ancestors to the most recent findings about child development, I'll draw a completely new picture of the origins of language. Fossils show that an evolutionary dilemma arose when our ancestors began walking on two legs. The narrowing of birth canals associated with upright walking made giving birth excruciating and dangerous. As so often happens, this dire situation was solved by an evolutionary balancing act: only the smaller, less-developed infants (and the mothers of these smaller infants) survived the ordeals of birth. Because of their physical immaturity, these newborns lacked the ability to cling unsupported to their mothers, a skill that monkey and ape infants very quickly develop. Before the invention of baby slings, women would have had little choice but to carry their helpless babies on their hips or in their arms. More important, they would have been forced to put their infants down as they gathered food.

When separated from their mothers, no doubt the babies fussed, as they do today, and busy prehistoric moms would have tried to soothe them. These mother-infant interactions began a sequence of events that led to our ancestors' earliest words and, later, the emergence of protolanguage.

Motherese may have influenced our young ancestors in other fascinating ways. Just as there are conflicting ideas regarding language origins, experts are also divided on when, how, and why music came about, and whether it had any evolutionary purpose. Some, like Steven Pinker, believe that music represents an entertaining but otherwise useless spin-off ("auditory cheesecake") from neurological machinery that evolved for different purposes. Other researchers reject the idea that music evolved from speech in favor of the reverse notion, as Darwin reasoned long ago. One thing is for sure: babies everywhere are extraordinarily musical, so it is not surprising that people sing lullabies and playsongs to them throughout the world. I believe that full-blown music and language evolved in lockstep with each other over millions of years of evolution as the musical and linguistic sides of the brain (right and left, respectively) gradually became larger and better at processing complex sounds.

Gesture may also have played a role, not only in the evolution of language but in the emergence of art. Modern children's artistic development parallels the startlingly early emergence and subsequent development of hominins' artistic expressions across the archaeological record. Like music, drawings and paintings seem to have emerged much longer ago than many researchers believe.

Certain mental processes, such as the ability to synthesize data, also develop as babies grow and contribute to the flowering of their verbal, musical, and artistic skills. Similar abilities must have evolved as our ancestors became linguistic, creative beings, and we'll see how changes in the brain facilitated the process.

This book draws heavily on what parents of small children see, hear, and do every day. These observations suggest that infants must have been of key importance during the extraordinary evolution of our species. Had Mother Nature not favored smaller, less-developed babies because of the terrible hardship of giving birth to bigger ones, our ancestors would never have invented motherese. And without motherese, our species' intellectual and artistic talents would not have blossomed. We would not have computers, the Web, or books with which to contemplate the mystery of our origins. In fact, we would not have developed into humans as we know ourselves today. But as we'll see, Mother Nature did favor helpless little babies, and the trials they faced radically affected their interactions with their mothers. The rest, as they say, is prehistory.

CHAPTER I

SILENCE
IS GOLDEN

RIDDLE

Who grasps the hairy tummy
of silent, knuckle-walking Mummy?
Hanging on, rides for free;
Baby is a chimpanzee!
—Grandma Dean

A LOT HAD TO happen before Neil Armstrong's historic journey aboard *Apollo 11* in 1969, and the world seems to have forgotten that someone else paved the way for his famous words, "One small step for man, one giant leap for mankind." Although the first individual who was spirited into space aboard the *Mercury-Redstone 2* in 1961 was a bit of a child prodigy, he wasn't just a narrowly focused space geek. He was a five-year-old chimpanzee named Ham.

Chimpanzees and humans share a common ancestor who lived some five million to seven million years ago, which is when Ham's ancestors and ours began to go their separate evolutionary ways—one by knuckle-walking on four limbs, the other by tottering along on two legs. That chimpanzees are our closest cousins is reflected in their appearance and behavior.

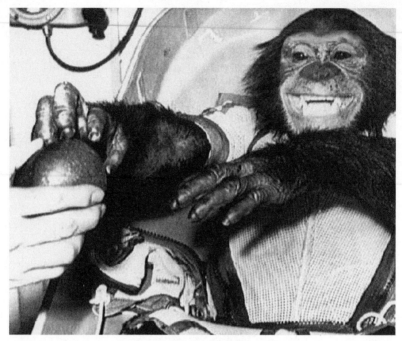

FIGURE 1.1. The first "small step" that eventually turned into a "giant leap for mankind" occurred millions of years ago when our earliest relatives began walking on two legs after splitting from space chimp Ham's early ancestors. *Photograph courtesy of NASA Johnson Space Center.*

They have large humanlike ears and expressive faces, and many of their emotional states and expressions are obviously similar to those of humans. Their body language is easy for humans to interpret because it so resembles our own. One of the two species of chimpanzees (*Pan troglodytes*, the common chimpanzee) hunts, makes tools, and makes war, while the other species (*Pan paniscus*, the bonobo) is famous for spending most of its time making love. And chimpanzees apparently have a sense of self. They are among a very few species that can recognize themselves in mirrors.

Despite the many behavioral similarities between chimpanzees and humans, observations of wild chimps reveal a dramatic difference: chimpanzee mothers treat their infants very differently from the way humans do. Although they are chatty and noisy by nature, when it comes to vocalizing to their young, chimpanzee moms are nearly silent compared to human mothers. Why should silence be golden for chimpanzee but not human mothers? To understand the subdued role of vocalizing in mother apes, we must first examine their experience of maternity.

CHIMPANZEE MOTHERS

The differences between chimp and human motherhood begin with birth. Giving birth is much easier for chimpanzees and the other two great apes, orangutans and gorillas, than it is for women. The chimp's gestation is shorter, about seven and a half months compared to our nine. Chimpanzee moms, like women, usually have one infant at a time, but because their newborns are much smaller than human babies, birthing is a quick, roomy, and smooth sail.[1] Of course, few primatologists have seen a chimpanzee delivery—they frequently occur at night, in solitude.[2] One primatologist, Frans de Waal, did manage to observe a midday chimpanzee birth at the Yerkes Regional Primate Research Center. The mother, Mai, stood half upright with her knees slightly apart, holding one hand between her legs. After about ten minutes, she tensed, squatted, and expelled the baby into both of her hands. She then took the baby to a corner, where she cleaned it and "consumed the afterbirth with gusto." De Waal was particularly impressed that, unlike monkeys, the other chimpanzees appeared interested and excited about watching one of their own clan give birth.[3]

Unlike roly-poly human babies, ape newborns are born "decidedly 'skinny' and horribly wrinkled."[4] But they resemble human newborns in their complete helplessness and vulnerability. Newborn apes depend entirely on their moms for suckling, warmth, comfort, protection, and transportation. Sustained physical contact between mothers and infants is vitally important. Unfortunately, although wild chimpanzee mothers tend to be intensively nurturing, a few are distressingly unloving toward their infants.

The Mommy Dearest Award for Bad Parenting goes to a chimpanzee named Pom, who was closely observed by Jane

Goodall. (Goodall began her groundbreaking research on wild chimpanzees in the 1960s at the Gombe National Park in Tanzania, and she has watched generations of them grow up ever since.[5] Much of what is known today about chimpanzees derives from her efforts.) Pom's mother, Passion, was a loner who became a cold, intolerant mother. She seldom played with Pom, and in turn, Pom was a clinging, anxious infant who seemed ever fearful that her mother would leave her behind. She had an especially traumatic time during weaning and was tense in the presence of males when she became sexually mature. Goodall contrasts Passion's parenting style with that of a more sociable chimp named Flo, mother of Fifi. Flo was a highly competent and affectionate mother, and Fifi enjoyed the additional advantage of having older siblings. Unlike Pom, she became a self-confident and assertive infant who tolerated the weaning process reasonably well and later grew to be relaxed in her sexual interactions. When Fifi became a mother, she was always alert to potential danger and often rescued her infant before it showed any signs of fear or distress.

Not so Pom, who was about thirteen years old when her first baby, Pan, was born. Unsurprisingly, Pom's behavior reflected the manner in which she had been raised:

> She found it difficult to cradle Pan comfortably when he was small—or else she simply couldn't be bothered. Often, as she sat in a tree, the infant would slip down off her lap and hang on frantically with wildly kicking legs as he tried to pull himself back up again. Only when he whimpered did Pom look down and, appearing slightly surprised, gather him back onto her thighs. But she seldom made any attempt to make a better lap.... Pom, like Passion, tended to move off without first gathering her infant.[6]

Before he reached the age of three, a fierce gust of wind swept Pan "like a stuffed toy" from a tree. He died within three days. This sad story underscores the importance of sustained physical contact between chimpanzee mothers and infants.

HANGING ON FOR DEAR LIFE

The danger and uncertainty of motherhood cause chimpanzees to be very possessive and protective of their newborns. Most chimp mothers keep their babies snuggled to their chests, where they contentedly nurse. The infant is helpless for the first few weeks of its life, so the mother must support it as she travels and feeds. This is a vulnerable time for both mother and baby, and Goodall calculates that nearly 30 percent of the newborns at Gombe die in their first year. Some of these deaths are even due to infanticide. No wonder new mothers are wary of letting others, even older siblings, get too close to their newborns.

In addition to needing protection, newborns slow mothers down. Sometimes primatologists don't see new mothers for one or two weeks after they give birth because they are unable to keep up with the rest of the group. They must use one hand to support their infants when moving along the ground, which can be awkward because knuckle-walking usually requires four limbs. When swinging through trees, mothers "make a lap" for their tiny babies by flexing their thighs. These postural adjustments for accommodating newborns probably explain why mothers cut down on travel. Because the foraging lifestyle of chimpanzees normally entails spending about half of each day feeding and much of the rest traveling to new areas, these first weeks can't be easy.

Just how difficult it is for new mothers to keep up was underscored by the unusual birth of twins in 1977 at Gombe to a chimpanzee named Melissa. When Gyre and Gimble were only a few days old, Melissa alternated traveling very slowly with stopping, sitting down, and cradling the twins. Within weeks, the infants were able to cling to their mother's abdomen: "But one kept clinging to the other by mistake: then he would pull his brother loose and both would start to fall, uttering loud cries of distress. Melissa had to give them almost constant support, holding them close with one arm, or travelling with her legs bent so as to support

FIGURE 1.2. Melissa with four-year-old daughter Gremlin, 1974. Ten-year-old Goblin is also pictured. *Photo by Curt Busse.*

their backs with her thighs. Once, one of the twins half-fell and banged his head on the ground."[7]

Unfortunately, little Gyre failed to thrive and died of pneumonia. Had Melissa been less protective and willing to share the twins with their big sister, Gremlin, Gyre might have lived.

Happily, in 1998 Gremlin herself gave birth to twins and, at this writing, Golden and Glitter have reached their tenth birthdays.[8] Although having twins interfered with Gremlin's ability to travel, Golden and Glitter are still alive today partly because Gremlin was not as reluctant to accept help from their big sister, Gaia, who was five and a half years old when they were born.

Because of the rapid motor development of their infants, most chimpanzee moms catch up quickly after childbirth. By the time they are two or three weeks old, chimpanzees can usually cling to their mothers for long periods without any support. Of course, human babies have lost the ability to cling tightly onto their traveling mothers, an evolutionary development that had profound implications.[9]

FIGURE 1.3. Young Freud riding on the back of his mother, Fifi, at Gombe. *Photo by Curt Busse.*

An infant chimpanzee clings to its mother all day as she travels in search of food and shares its mother's sleeping nest at night. At first the tiny infant remains attached to its mother's chest during travel, but as it grows heavier it shifts to her back. By the time the baby is three to five months old, the mother allows other youngsters to approach and play with it. Of course, Mom still keeps a wary eye out and rushes to defend her infant from older offspring or others if they are too rough.

A mother is more likely to allow siblings to help carry the baby after it takes its first steps—usually by six months of age. But most mothers remain quite protective of their young in social situations. They are often reluctant to allow even other "aunts" to hold or carry their infants. Before she gave birth to twins, Gremlin had a son named Getty. Goodall recounts that Getty was ten months old before Gremlin allowed him to be groomed briefly by her own mother, Melissa. On one occasion, Melissa sat with her grandson on her lap and groomed him. Gremlin cautiously glanced into her mother's face, reached for her son with a soft, pleading whimper, and took him away. According to Goodall, Gremlin feared that her mother might try to steal her beloved son.[10]

Maternal possessiveness is adaptive: only well-protected infants stand a decent chance of living to an old age and becoming parents themselves. This is vividly illustrated by Goodall's documentation of incompetent, off-guard, or injured mothers who lost their babies to infanticide and cannibalism. Passion, the cold mother discussed above, and her late-adolescent daughter, Pom, were in the habit of seizing infants from their mothers and killing them by biting into their foreheads. They would then eat the infant, sometimes in front of its mother. Although Goodall initially assumed that the infanticide and cannibalism of Passion and Pom were pathological behaviors, newer research suggests that high-ranking chimpanzee females like Passion and Pom attempt to kill infants as an unconscious strategy for removing individuals who might compete with their own offspring.[11]

Male common chimpanzees also pose a threat because they sometimes attack, kill, and cannibalize infants; these babies usually belong to mothers who have wandered away from other communities. The male predilection for violence, however, goes beyond the aggression and infanticide shown by Passion and Pom. Groups of adolescent and adult males (accompanied occasionally by a female who is sexually receptive) regularly patrol the edges of their territories seeking opportunities for fights with enemies from neighboring communities. Such aggressive territoriality can be devastating: one chimpanzee community at Gombe, the Kahama, was completely annihilated in a series of attacks by males from the Kasakela, another community.

Chimpanzees are preyed upon not only by their own kind but also by leopards, predatory birds, and humans, which probably has something to do with why young infants cling tenaciously to their mothers and why mothers are reluctant to let others hold them. As noted by renowned anthropologist Sarah Blaffer Hrdy, "Chimpanzee mothers have a special worry. These apes are unusual among primates insofar as they hunt, and mothers have to worry about the possibility of a meat-hungry chimp eating her newborn. It's probably significant that chimps are more carnivorous than most primates."[12] Indeed, hunting and meat eating were very important during human evolution, so our ancestral mothers probably would have shared this special worry.

Bonobo mothers are every bit as protective of their infants as common chimpanzee mothers, and their babies travel glued to their bodies in the same way. This protectiveness comes even though bonobo males have not been reported to patrol their borders or commit infanticide and generally seem to be more gregarious and less aggressive than common chimpanzee males.[13] Unlike common chimpanzees, bonobo society is more egalitarian when it comes to sexual politics, because female bonobos are much more sexually receptive than other female chimpanzees. Interestingly, some anthropologists have speculated that the extreme sexuality of female bonobos is actually a way of preventing infanticide, because males are less likely to kill infants who could be their own offspring.

Protection isn't the only reason why mothers and babies cling to one another. As with humans, maintenance of close physical contact between infants and mothers is intimately linked to nursing. Although it is crucial for common chimpanzee and bonobo infants to travel attached to their mothers, they cannot do so after a certain age. A three-year-old chimpanzee continues to suckle and ride on its mother's back, but during the next two years the infant is gradually weaned and forced to walk on its own. (Bonobos nurse until infants are about four years old.) The weaning period is very difficult for a juvenile, who until then has enjoyed round-the-clock access not only to the mother's milk but also the warmth and reassurance of her body. Infants resist being weaned, so mothers are sometimes aggressive to get the job done. They can also be extremely tolerant of their subdued and depressed offspring. Even coldhearted Passion tried to soften the blow when little Pom was being weaned: "She almost always responded to Pom's requests for grooming and even allowed her to ride on her back with a minimum of protest. For weeks after we were sure that her milk had dried up, she let Pom sit close, a nipple in her mouth, her eyes often closed, for as long as twenty minutes at a time."[14]

Infants manipulate their mothers during these difficult times and frequently throw temper tantrums. When this happens, their mothers often embrace them and allow them to suckle. For example, after a four-year-old common chimpanzee was twice rejected while attempting to climb onto his mother's back, he uttered terrified screams that galvanized the mother into action. She "rushed back and with a wide grin of fear, gathered up her child, and set off—carrying him."[15] Similar dynamics occur when bonobos are weaned: "The tantrum is easily dispelled by embracing the mother and clinging to her nipple. . . . The effectiveness of the mother's nipple is striking."[16]

It takes a mother's dogged persistence to accomplish weaning in the face of such resistance, and sometimes it doesn't work. According to Goodall, an aged female named Flo had been a highly successful mother, but she was too old and weak to handle baby Flint's assertive temper tantrums during weaning. Flint nursed up until his little sister, Flame, was born. Although he was then weaned from the breast, he continued to ride on his mother's back and even attempted to ride on her abdomen like an infant by attaching himself over the body of his sister. Flint became increasingly depressed and spent hours grooming his mother. He lightened up, however, after six-month-old Flame died. Flint continued to ride on his mother's back until he was eight years old. When old Flo died shortly after that, Flint was bereft. Having apparently lost the will to live, he curled up and died near the spot where his mother's dead body had been found.[17]

Although Flint may have been unusually attached to his mother, common chimpanzee orphans do not generally fare well. Orphans who are less than three years old almost always fail to survive, but nutritionally independent orphans between four and six years of age may also die, even if they are adopted by older siblings. These infants are extremely depressed after losing their moms, and it is simply heartrending to read about the valiant attempts of other chimpanzees to care for them. Goodall, for example, marveled at the care and concern shown by a twelve-year-old adolescent male, Spindle, when he adopted a three-year-old unrelated orphan, Mel.[18] Spindle shared his night nest and food with Mel, and protected him in social situations. When Mel whimpered during travel, Spindle carried him on his back or allowed him to cling to his abdomen in the infantile position. Spindle carried Mel so much, in fact, that he developed bald spots where Mel's feet gripped his hair. Thanks to Spindle's tender care, Mel managed to survive his mother's death. Of course, being allowed to ride on Spindle's back was particularly vital. Sometime during evolution, human infants lost the ability to cling unsupported while traveling on their mothers. Later, we'll see how this seemingly simple change had far-reaching repercussions on the direction of human evolution.

CHIMPANZEE GESTURES AND VOCALIZATIONS

In the study of language origins, much of the relevant research on apes has shifted from the field to the laboratory or to the rare home in which an ape has been raised by humans. Because of these more artificial studies we now know, for example, that all three great apes are capable of learning very rudimentary forms of sign language that deaf humans use. We must keep in mind, however, that apes raised by humans are far removed from the experiences that would have shaped their thoughts and actions in the wild.

Because chimpanzees are genetically and behaviorally closest to humans, they are of special interest when speculating about what kinds of communication existed before speech evolved. Chimpanzees are highly communicative, both in gestures and sounds. They frequently use gestures to express moods, call attention to themselves, or make demands. Unlike that of humans, the body language of chimpanzees is almost always egocentric, rather than conveying information about external events, objects, or "third parties."[19] Furthermore, chimpanzee gestures are highly likely to involve physical contact with another individual. Many of chimpanzees' emotional states are also expressed in a variety of easily recognizable facial expressions. These in turn are frequently linked to particular vocalizations.

Although there are vast differences between the vocalizations of apes and humans, Goodall showed that the vocal communication of common chimpanzees is far more complex than previously

appreciated. She classified thirty-four calls along with their associated emotions, observing that one type of call glides into another.[20] For example, single *hoos* may be uttered several times in
succession, but as the sequence speeds up and starts to rise and
fall in pitch and volume, it grades into a *whimper. Squeaks, screams,*
and *pant grunts* are frequently used as submissive vocalizations;
hoots, barks, and *coughs* are likely to be nonsubmissive or even aggressive. Interestingly, chimpanzees can identify unseen individuals by their pant hoots, pant grunts, and screams—even when
scientists cannot.[21] Despite this, males tend to interrupt each
other's vocalizations with their own, which hampers the ex-
change of information.

Unfortunately, few field or laboratory studies have focused on
the communication between mothers and infants. One study,
however, that occurred in Moscow over ninety years ago is highly
relevant in this regard. Nadia Kohts (or, more fully, Nadezhda
Nikolaevna Ladygina-Kohts, 1889–1963) was a comparative psychologist who carried out a detailed study of a common chimpanzee named Joni that she raised in her home from 1913 to
1916. After her own son, Roody, was born in 1925, Kohts began a
comparative study of development in Joni and Roody.[22] Although
Kohts's goal was to test Joni's perceptual and conceptual abilities,
her research provides a wealth of information about Joni's need to
be nurtured by a mother figure that is consistent with what more
recent studies have provided. Lacking the benefit of subsequent
research by Goodall and others, Kohts used corporal and psychological punishment to control Joni. Nevertheless, Kohts became
so deeply attached to Joni that she wrote "in my heart, both of
them, Joni and Roody, take up an almost equal space."[23]

It was clear from the day that Joni first arrived at the Kohts
home that he obtained "contact comfort" from the rags in his
cage, just as baby chimps obtain comfort from their mothers:

> Tampering with his sheets, which was necessary for their
> daily airing, for a long period of time had been the reason for
> his angry protests and violent defiance. He became totally
> unresponsive to all requests, persuasions, yells, and even cor-

poral punishment; he would not give up his rags for the world. He gripped them, holding them as fast as he could; he did not let them go even when he was dragged with them. He held them not only with his hands, but also with his teeth, and he pressed them to his body if they were about to slip away from him.[24]

Joni was only about one and a half years old when he arrived in the Kohts household and would not have survived as an orphan in the wild. He readily adopted Nadia Kohts as his mother, and the two bonded. Sadly, Joni was locked in a cage at night, which he hated. His strongest desire seemed to be for "tactile contact with the living creature who presents him with joy":

> When Joni wants to eat, drink, or sleep, he literally does not take his eyes off me; he follows me everywhere, staring at my face all the time. He has developed a special conditioned reflex for the expression of his desire to drink: He runs up to me and sucks at different open parts of my body (at the hands, neck, or face), and every time he greedily drinks the water that I offer him in response to his expressive prose. . . . He willingly falls asleep on my lap and can sleep a long time in such a position. He would be even more willing to sleep in a bed with me and protests when I do not permit this.[25]

Joni was also sensitive to his adopted mother's reprimands and would whimper and cry even if she simply waved a hand to indicate that he had not performed well on a test. If she was slow to comfort him, Joni extended his arms and begged to be taken onto her lap. If she refused, he cried violently and took a long time to recover, even after she comforted him. Kohts viewed the face as "the mirror of the soul" and described and photographed similar facial expressions in Roody and Joni for laughter, crying, fear, anger, surprise, attention, revulsion, and general excitability. She also studied their body language and sounds: "In the child and chimpanzee, . . . the expressiveness of the bodily manifestations in gestures and sounds corresponds with the strength of emotions."[26] For example, in fits of despair, both chimpanzees and children may place their hands on their heads and batter their own bodies with their fists.

Kohts found that gestures and images played a more important role than words in training Joni, although Joni clearly understood a variety of Russian imperatives. Nevertheless, "an eloquent language of instinctive sounds" accompanied Joni's emotionally expressive gestures and body movements. Significantly, Kohts noted that Joni was more likely to vocalize when sad than happy: "Sad or unpleasant feelings of the chimpanzee are accompanied by diverse and extremely strong sounds; in contrast, pleasant and joyful emotions are almost silent. You rarely can observe a sad chimpanzee extending his lips in discontent without some groaning, but you often can see the silent smile of a joyous animal."[27] Sad chimpanzees cry but do not shed tears the way human infants do; happy ones may laugh with a wide-open mouth accompanied by rapid breathing. Kohts also concluded (correctly) that chimpanzees like to make sounds apart from those they produce vocally. Joni expressed both positive and negative emotions by knocking, banging, slamming, slapping, tipping over, hurling, and breaking objects. Although Joni could be extremely noisy, his adopted mother had not noticed "any attempts on his part to reproduce or imitate even a semblance of intelligible human sounds."[28] Kohts's observations were truly remarkable in her day, and many of them have since been confirmed and extended in larger samples of chimpanzees.[29]

MOTHER-INFANT COMMUNICATION: A TWO-WAY STREET

It is clear from Kohts's interactions with Joni that he was able to communicate his emotions and desires to her, and this is equally true of chimpanzee infants in natural settings. Happily, the communications of the latter are not as sad as Joni's because chimpanzee mothers remain in constant physical contact with their small infants. By the time they are about three months old, infants watch their mothers' faces intently (recall Kohts's remarks about Joni staring at her face) and begin to communicate with facial expressions. Before reaching six months of age, for example, they request play or tickling from their mothers with *play faces*. Even before that, chimpanzees ask for solid food by putting their mouths near their mothers' mouths, whereas bonobos usually just touch their mothers' mouths. Other infants may simply grab the food they want. As they continue to mature, youngsters communicate their emotions and desires to their mothers with a variety of gestures, including the dramatic temper tantrums that occur around the time of weaning.

Chimpanzee and bonobo mothers use a rich repertoire of facial expressions and body language to communicate with their young. In the wild, these occur in various contexts related to carrying, cradling, nursing, weaning, play, traveling, and learning motor skills.[30] Captive chimpanzee mothers have been observed examining and cradling their infants, and kissing them on the mouth. They may also pat their infants' heads and encourage their development by giving them walking lessons.[31] Mothers also lend a helpful hand or foot to more mature youngsters that are moving or playing in trees. Some of the most interesting gestures are used in conjunction with feeding. Mothers snatch leaves from their infants' mouths if they are not part of a normal diet. Goodall believes this sort of intervention reinforces "cultural" food preferences in different chimpanzee communities. Along these lines, chimpanzee mothers from the Taï forest have anecdotally been reported to teach their offspring to crack nuts open with rocks.[32]

Play is the highlight of a young chimpanzee's life, occurring most often in youngsters between the ages of two and four. Females with infants play more than other adults: "A chimpanzee infant has his first experience of social play from his mother as, very gently, she tickles him with her fingers or with little nibbling, nuzzling movements of her jaws. Initially these bouts are brief, but by the time the infant is six months old and begins to respond to her with play face and laughing, the bouts become longer. Mother-offspring play is common throughout infancy."[33]

Bonobo mothers play with their infants using slow-moving, gentle movements, often while resting. Mothers tickle, play bite, and grab their infants. "While lying sprawled looking up, she will tickle the infant and hold its hands and feet; hanging high in space, the infant looks very happy and fortunate."[34] These mothers sometimes appear to be playing the human game of "airplane" with their infants. Play biting also occurs in common chimpanzees and may entail taking turns: "The early biting triggered the onset of mother-baby play: contingent upon when bitten, the mother started to tickle the baby and this biting-tickling grew into an alternating interaction in which both mother and baby could take their turns."[35] Learning to take turns, of course, must occur for language to develop.

In the next chapter, we'll see that human mothers (and fathers) speak to their babies in a special tone of voice known as baby talk or motherese that, among other functions, helps infants learn about language. Unlike chimpanzee mothers, human parents also encourage their infants to vocalize. This raises a profoundly important evolutionary question: "The chimpanzee is usually very quiet, unless it is emotionally aroused. The spontaneity of noncrying vocalization is usually low. Does a chimpanzee mother-infant pair interact vocally?"[36]

If a youngster wants to ride on its mother's back it may utter a *hoo*. Babies also produce screams if they are threatened or hurt, and *whimpers* or *tantrum screams* are especially prevalent during weaning. Smaller infants are quieter. If a baby that is hanging onto its mother's underside begins to fall off, it may whimper. For the most part, however, infants seem happy to ride along in silence, which also seems to be the case when they grow large enough to ride piggyback. Similarly, there are few reports of chimpanzee and bonobo mothers directing vocalizations specifically toward their infants. To be sure, mothers will reply to the screams of their infants, even if they are out of sight. They also produce *soft barks* or *coughs* as mild rebukes toward weaning youngsters, and they may make soft vocalizations while examining their babies. Below is a rare report of mother-infant vocal interactions, which Goodall observed under unusual conditions: "Poor Melissa—the crying of one sick twin was bad enough, but so often Gimble joined in, frightened, perhaps, by the intensity of his brother's calls. Sometimes when they yelled Melissa sat and cradled them until they quietened. But at other times, holding them tightly, she moved on very fast, uttering a series of cough-like grunts—as though she was threatening them."[37]

One of the few circumstances under which mothers routinely vocalize to their infants is during traveling and foraging. Both bonobo and chimpanzee mothers utter *hoos* to retrieve their infants for travel, and "soft grunts may be exchanged when two or more familiar chimpanzees, especially family members, are foraging or traveling together. Typically one individual grunts when he pauses during travel, or when he gets up to move on.... Thus these grunts function to regulate movement and cohesion."[38] Significantly, on those rare occasions when mothers vocalize to their infants, their utterances are similar to the vocalizations their infants produce.[39]

It is clear that chimpanzee infants scream, squeak, and whimper under aversive conditions and that their mothers hear and respond to them. Youngsters may also produce short, breathy *effort grunts* when struggling. But what about vocalizations under less trying circumstances? Laboratory studies at the Primate Research Institute in Kyoto, Japan, showed that a day-old chimpanzee named Pan responded to loud noises with *staccatos* and *grunts*, and at two months old produced similar vocalizations in response to various stimuli, including being spoken to.[40] By ten weeks, she laughed and imitated some environmental sounds. Pan, who had been brought up by humans, eventually gave birth to a daughter, Pal, whom she raised. As was the case for Pan, Pal's vocalizations were almost always in response to outside stimuli, including the vocalizations of her mother or humans, the presence of other chimpanzees, or environmental sounds. Significantly, both chimpanzees decreased their vocalizations as they matured, apparently because their surroundings had become more familiar.

In an eloquent study, Shozo Kojima compared vocal interactions between three pairs of chimpanzee mothers and infants. Of the three mothers, only one (Pan) had received vocal training from humans. The other two mother-infant pairs did not interact much vocally. However, Pan and Pal did, particularly in the evening. Pan reacted to Pal's unhappy squeaks, whimpers, and screams with cordial pants, pant grunts, or soft grunts. Kojima attributed the unusual amount of vocal interaction between Pan and Pal to Pan's early vocal training from humans.

Kojima points out that the vocalizations of human babies initially develop similarly to a chimpanzee's, but with one big difference: in addition to responding to external stimuli, the human infant produces many spontaneous vocalizations, which, over time, lead to babbling. And babbling, of course, is a warm-up for the first appearance of speech. Ultimately, it is clear that chimpanzee mothers do not engage in humanlike motherese, and their infants never learn to babble, let alone talk. For a chimpanzee, vocalization is limited. For a human, it clearly is not. But how did this difference evolve, and what effects did it have on our subsequent evolution as a species?

CHAPTER 2

SAYING
WHILE
SOOTHING

A MOTHER
WAS PRESSING HER BABE

A mother was pressing
Her babe to her breast
And saying while soothing
His sorrow to rest,
Sleep gently my darling,
Sleep soundly my boy
For thou art my treasure,
My rapture and joy.
—*Welsh lullaby*

CONTEMPORARY CULTURES THAT MOST resemble those of our early ancestors can give us clues about the evolution of mothering. Because industrialization has transformed much of humanity, anthropologists traditionally have speculated about human evolution by focusing on the dwindling non-industrialized (or "traditional") societies. In this chapter, we will do the same, keeping in mind that although the lifestyles of traditional societies resemble those of our early ancestors, it would be a mistake to consider them equal.

For example, modern hunting and gathering societies rely on many modern innovations: carrying devices, food-sharing between group members, and camps, home bases, and settlements. They also have modern tools and technologies such as traps, bows and arrows, nets, snares, cooking pots, and cloth.[1] As Frank Marlowe, an evolutionary anthropologist at Florida State University, has explained, we must ignore the effects such technology has on modern hunters and gatherers if we are to compare their lifestyles to those of our ancestors.

We will therefore focus heavily on various nonindustrialized cultures to discover the essence of contemporary maternity, child-rearing practices, and mother-infant communication. What is universal? What is not? Do women share any maternal behaviors with chimpanzees? More to the point, can traditional societies shed light on the emergence of vocalization between mothers and infants that set our species apart?

HUMAN MOTHERING

Ape mothers breeze through childbirth because the heads and shoulders of their infants are so small. But for the human mother, the process can be excruciating. I know this firsthand, having experienced unmedicated childbirths for both of my daughters. Though human newborns are pudgier than those of apes, the tight fit during human deliveries does not occur because our fetuses are larger relative to their mothers or, contrary to common wisdom, because they have unusually big heads.[2] Labor is difficult for women because the shape of our ancestors' pelvises changed to accommodate the rearrangement of the muscles used for upright walking, which caused the birth canal to get smaller.[3]

Not just women have a difficult time with birth. The trip down the birth canal may be hazardous for their babies as well. Sarah Blaffer Hrdy says:

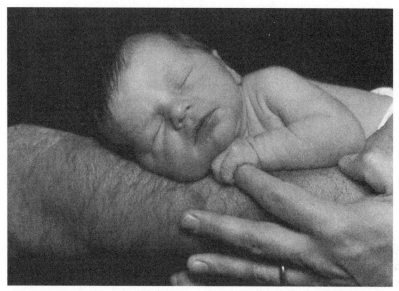

FIGURE 2.1. *The author's grandson, Jacob Riddle, soon after he was born in 2002.* Notice his baby fat, little grasping hand, and how content he seems to be snuggled against his father's hairy arm. *Photo by Michael Riddle.*

> Birth is a perilous time, when the alien, ghostly world of fetuses comes in contact with the human world. If things go badly during passage of the large-headed baby through the birth canal, such a ghost might easily claim the mother's life. It is the neonate that mediates between these worlds and must remain in limbo after birth till rendered safe.[4]

A study of several thousand births before modern obstetrics developed reported that about one baby in twenty died during its first month of life.[5] High infant mortality still plagues non-industrialized communities.[6] Given the rigors of navigating tight birth canals, anthropologists have long been puzzled about why human fetuses put on a substantial layer of fat—up to 16 percent of their birth weights—before being born.[7]

To account for this seeming oddity, Hrdy discusses four hypotheses.[8] The "insulation" hypothesis says baby fat may have developed as protection against nighttime cold when early hominins first began sleeping on the ground. A second hypothesis is that stores of fat might be "insurance policies" if the baby is orphaned or its mother cannot nurse—but then, as Hrdy notes, why wouldn't this be the case for other primates? It is interesting that women's milk supplies take a few days to build up after they give birth and that babies tend to lose weight after they are born but generally put it back on by around two weeks of age.[9]

The "food-for-thought" hypothesis suggests that neonatal fat is a metabolic stockpile for fueling the amazingly rapid brain growth that occurs during an infant's first year of life.[10] Finally, assuming that chubby babies are more likely to live, Hrdy proposes a fourth hypothesis, "self-advertising," where mothers are more likely to perceive fatter babies as healthy and therefore "keepers." Hrdy reminds us that parents in our own culture proudly report birth weights on their birth announcements.[11] Because various traditional societies have practiced infanticide in the past (and some still may), this hypothesis should not be viewed as trivial.

All four of these hypotheses may well have elements of truth. A complementary possibility is that since human babies fatten up right before birth, they may be putting on padding to cushion them a bit as they squeeze through the birth canal, which is bordered by the rigid edges of the mother's bony pelvis and the curved point of her tailbone. Full-term fetuses press their chins down during delivery, which increases the exposure of their necks and shoulders, which could explain why prenatal fat builds up on the backs of newborns' necks and between their shoulder blades. The dimensions of infants' heads are unlikely to be affected much, if at all, by the amount of baby fat, so their mothers' cervixes must dilate to certain sizes regardless of whether the infants are butterballs. Although an added layer of fat might make infants' deliveries a bit more snug, it could also make them smoother. If you had to travel through a tunnel with bony edges and protuberances and you could choose to do so either in a well-padded suit that slightly increased the squeeze factor or naked but with a bit less pressure from the tunnel walls, which would you choose? I know what my answer would be.

CHILDBIRTH AROUND THE WORLD

Because childbirth is so difficult for women, *Homo sapiens* is the only primate that routinely relies on attendants to help mothers during labor and delivery.[12] Unlike other primates, however, human norms for childbirth and child rearing vary considerably and are constrained by local cultural taboos, spiritual concerns, and religious practices.[13]

However, women in small traditional societies may rely little, if at all, on others' help during childbirth, as illustrated by the Ifaluk people, who live on two tiny coral islands in Micronesia. As of 1995, their population included only about six hundred individuals who made their living mostly by fishing, cultivating taro root, and harvesting breadfruits and coconuts.[14] Ifaluk women traditionally gave birth in special huts and were attended only by female relatives; today, midwives may also attend. But these attendants merely observed the births, although they buried the placentas and helped care for the babies after they were born. Mothers were encouraged to extract their own babies and to refrain from crying out so as not to shame themselves or their families.

Giving birth in kneeling, sitting, or squatting positions with the help of one or more attendants remains the standard practice in many traditional societies. For example, laboring Beng women, who live in small farming villages in the West African nation of Ivory Coast, sit on the floor with their legs outstretched and lean back supported by female family members.[15] Some Muslim women who live in small farming villages in Turkey also deliver while seated, with the help of their mothers-in-law (who rub their backs) and a midwife (who may massage their abdomens).[16] In this case, the mothers-to-be sit on low stools inside washtubs. When pushing, they extend their legs and press them against the sides of the tub, which they also grip with their hands. Between contractions, women may walk around and even eat. Traditional Hindu women from Bali squat on new mats in their household compounds and give birth with the help of a midwife and in the

company of their husbands and other kin, contrary to the taboo
that many cultures have against men being present.[17]

In earlier times, husbands never would have been present during "borning" among the seminomadic Warlpiri, who lived until
recently as hunters and gatherers in the central Australian desert
and for whom birth traditionally has been private.[18] Frequently,
only one woman, ideally the mother-to-be's grandmother, attended and helped her through delivery. If the group was on a
long hunting or gathering trip when a woman went into labor,
she and her helper would stop for childbirth and then catch up
with the group later. Sophia Pierroutsakos of Furman University,
South Carolina notes:

> If a woman was giving birth in the bush, the midwife would
> help her into a squatting position. By supporting her shoulders
> and rubbing her belly, she would facilitate delivery. As the preg-
> nant woman squatted, the baby would be delivered directly
> onto the soil. The land is our mother and provides for all our
> needs, so it is important that a new life should arrive directly
> onto her surface.[19]

The helper also built a fire "that provided warmth and kept
flies away as the woman lay naked." After the birth, either the
midwife or the mother buried the placenta at the edge of camp,
and the baby was "smoked"—held over smoking acacia leaves—
to give it strength.

People among the cattle-herding Fulani of West Africa who
live in small villages or semipermanent camps also believe it is
important for newborn babies to come into contact with the
ground to establish a powerful connection between them and
their new homes.[20] Like the Ifaluk, laboring Fulani women traditionally have been advised to take responsibility for their own deliveries and to avoid making too much noise in the process:

> Be prepared to give birth alone or with a companion. When
> you feel the pains come, which we call *luuwa*, the "struggle"—
> light a fire and squat on the floor of your shelter. As you feel
> the pains intensify, grit your teeth, close your eyes, and take

up the kneeling position. Try to keep calm and do everything you can to avoid screaming. It is shameful to show fear of childbirth, and if your co-wives and mother-in-law hear you, you will never hear the end of it.[21]

The new mother's female companions enter her shelter only after they hear the newborn's cry. They then help to cut the umbilical cord and bury the afterbirth where the newborn first touched the ground.

Of course, these birth practices represent only a smattering of those used by women in nonindustrialized communities around the world. If they suggest a common theme, it is that most cultures recognize that it is better for unmedicated women to deliver in upright postures than on their backs, which makes sense because of gravity. (This is also the normal posture for other primates.) There are additional similarities in how cultures treat their new arrivals. Immediately after birth, newborns are thoroughly inspected and their umbilical cords severed. The afterbirth is disposed of, with or without ceremony, frequently by burying. Soon after their appearances, newborns receive their first of (usually) many daily baths, often by a grandmother or midwife. In traditional cultures, the age at which infants normally are weaned vary from one to four and a half years (with a median age of two and a half years).[22] The timing and extent to which other foods are introduced also varies. Infants in these communities usually sleep beside their mothers, sometimes for years. Most infants in nonindustrialized societies also seem to be highly valued—even prized.

Most, but not all. Depending on the culture and the circumstances surrounding a particular birth, a mother (or her husband) may not welcome another mouth to feed. Factors such as having twins, bearing a child of a particular sex, already having one young infant, or giving birth to an infant with deformities sometimes tilt the balance toward benign neglect, abandonment, or even infanticide.[23] These behaviors also occur in industrialized societies, of course, although with less frequency, partly because abortion is more readily available. In fact, Hrdy believes that parent-inflicted infanticide has been a human practice since prehistoric times, in contrast to the killing of others' offspring that some nonhuman primates, including chimpanzees, practice.[24] The observations of infanticide that anthropologists have recorded are heartrending, but they become more understandable when viewed within cultural contexts. For example, in some parts of the world, infanticide optimizes the chances of an infant's older sibling having enough breast milk to survive. This is by no means an inconsequential concern because of the high childhood mortality that plagues nonindustrial populations.

Since nonhuman primates rarely kill their own offspring, it is interesting to speculate, as Hrdy does, about when, how, and why humans started this practice. Whatever its causes, however, infanticide occurs most when infants are perceived as unlikely to survive. Recorded rates of infanticide range from nearly zero in many traditional African cultures to more than 40 percent during the 1970s among the Eipo horticulturalists of Papua, New Guinea, who preferred male offspring.[25] (After missionaries arrived in the area, the figure dropped to 10 percent.) Cross-cultural studies show that the most common causes of parent-inflicted infanticide are lack of paternal support, unfavorable environmental circumstances, and poor infant health.[26] If food is scarce or a mother is currently nursing another infant, having to care for a newborn could threaten survival for older siblings and perhaps even the mother. If a newborn is deformed, the energy required to raise it might better be used to give birth later to another, healthier infant. Viewed this way, parent-inflicted infanticide arguably serves, on average, to increase other family members' chances of survival.

Despite the grim prospects for some newborns, most infants are held dear in traditional communities and are often lavished with affection and patience. Because babies are such a precious commodity and because infant mortality runs high, there is a widespread anxiety that healthy infants might fail to thrive. This concern has provoked an astonishing variety of traditional "treatments" for newborns and infants around the world that, at first, seem strange from a Western perspective. Sweating an Ifaluk infant is thought to make it grow, painting the soft spot of a Beng baby's head orange prevents illness, and rolling Fulani infants in cow dung makes them less attractive to witches and others who might want to steal them. Aboriginal Warlpiri mothers coat their infants with acacia ashes to prevent sunburn and hold them over burning leaves to give them strength. Beng and many other infants wear jewelry, not just for decoration, but also to ward off disease and prevent harm from supernatural forces.[27] Lest these practices seem odd, keep in mind that newborns in American hospitals are typically jabbed for blood samples, diapered, wrapped in blankets, and decorated with identification anklets and pink or blue beanies.

The concern about infant mortality is so deeply engrained that many traditional communities delay acknowledging the onset of personhood in newborns. Traditional Fulani are not regarded as persons until they are named in a ceremony on their eighth day, and the Ifaluk are first regarded as complete persons at ten days of age.[28] The Ifaluk also have a taboo against speaking directly about a newborn before it is several days old and considered healthy. The Beng believe an infant does not begin to become a person until its umbilical cord falls off, and then the transformation is not complete for six to seven years.[29] For Balinese, an infant becomes fully human at a special ceremony on its 210th day of life.[30] But the process doesn't need to take so long. An infant in a Turkish village becomes a person the moment he or she is washed, swaddled, and presented to the paternal grandfather, who whispers verses from the Koran into the baby's ear and bestows its name.[31] Most of these practices allow parents to suspend making an emotional commitment to newborns who might die. The advice to Beng mothers says it all: "If your child dies, it may comfort you to remember that the younger the baby, the more the little one was still living in *wrugbe* [the land of the dead]. Indeed, if the umbilical cord hadn't fallen off yet, the baby hadn't even begun to leave the land of the ancestors, and the village chief won't announce a funeral."[32]

HANGING ON—AND LETTING GO

Sustained physical contact with mothers is crucial for the physical and psychological health of young chimpanzees. By three weeks old, they can maintain the contact they need by grasping their mothers' undersides with their hands and feet. Our ancestors' infants, on the other hand, lost the ability to cling to their mothers' hairy bellies when pelvises adapted to upright walking. Because newborns began traveling through tighter birth canals, the ones with smaller heads—and therefore smaller brains—were favored by natural selection. So *Homo sapiens* babies are born small-brained and extremely immature compared to adult states—so much so, in fact, that they never develop the chimpanzees' ability to cling to their mothers, though they retain a grasping reflex. How is a human mother, or her poor baby, to maintain the physical contact necessary for survival?

Babies have many ways to ensure this sustained contact. Their desire for close physical contact with their mothers is every bit as strong as Joni the chimpanzee's. According to attachment theory, which British psychoanalyst John Bowlby pioneered, toddlers will go to extraordinary lengths to prevent separation from their mothers—kicking, screaming, crying, hanging onto a leg—or to reestablish contact.[33] Bowlby's evolutionary perspective recognized that infantile behaviors (or attributes such as cuteness) that promote parental attention would have given their tiny possessors an advantage in the survival sweepstakes. A human newborn, of course, is not yet developed enough to fling her arms, wailing, around her mother's leg. However, she can cry, and from the moment of birth, babies do just that to object to being physically separated from their mothers.

The pitch of the human infant's cries for attention rises and falls, resembling the calls of monkeys who are separated from their mothers. Crying in human infants, however, differs in its rhythm of short, breathy expirations, alternating with long intakes of air.[34] Within the first weeks of life, humans increase the amount they cry, peaking at about six weeks of age. To the extent that they cry at all, so do chimpanzees.[35] Their crying gradually levels off by the time they are three to four months old. This pattern appears to be the norm for newborns in a wide variety of cultures, as does the tendency for their cries to cluster in the late afternoon and early evening.[36] Unlike chimpanzees, our babies also develop the ability to shed emotional tears. By the time they are three months old, human infants have begun to modulate their cries to express anger, pain, and frustration. Consistent with anthropologist Meredith Small's suggestion that, like babbling, crying may be a precursor to language, their crying also becomes more interactive and intentional.[37]

Various studies have shown that nonhuman primate mothers recognize their own infants' vocalizations, so it is not surprising that women can distinguish their babies' different cries that indicate hunger, pain, or anger. But besides a reading of their babies' general moods, what are they picking up from these cries? According to Joseph Soltis, who has intensively studied crying in very young infants, crying may provide parents with a surprising amount of information. The cries of extremely sick infants, for example, are very high-pitched and sometimes associated with other acoustic features that convey acute and chronic distress. When questioned, parents characterize these cries variously as "aversive, urgent, arousing, grating, piercing, distressing, saddening, or sick."[38] Although some studies show that abnormal cries may lead to positive responses from parents, others reveal that high-pitched crying provokes strong negative emotions in some adults and may even place ill babies at risk for neglect, abuse, or infanticide. This sheds some light on the fact that infanticide usually occurs in nonindustrial communities when infants are perceived as being unlikely to survive.

This type of pathological crying differs from colic, which is sustained and inconsolable crying that occurs spontaneously and without any obvious cause. Acoustically, the cries of colicky babies are within the normal range, as is the development of the amount they cry each day, which peaks at about six weeks of age and then levels off and is resolved by the fourth month. For these reasons, Soltis views colicky infants as "extreme cases of an otherwise normal and universal increase in crying during the first few months of life."[39] Taking an evolutionary approach, he even goes so far as to suggest that colicky crying may have evolved to indicate that infants are healthy ("vigorous") and are therefore keepers. Soltis realizes that this is a controversial suggestion, but it may have an element of truth, as indicated by an Eipo mother who had planned to abandon her infant if it turned out to be female. After giving birth to an obviously healthy baby girl, the mother sat thoughtfully watching the infant squirming and screaming vigorously. Eventually, the mother abandoned the baby. Ambivalent, she returned two hours later to retrieve her daughter, who was too strong to abandon, she explained, almost apologetically.[40]

Small believes that crying and parents' sensitivity to it are adaptive traits that "evolved to serve the infant's purposes: to assure protection, adequate feeding, and nurturing for an organism that cannot care for itself. By definition, crying is designed to elicit a response, to activate emotions, to play on the empathy of another. . . . The caretaker has also evolved the sensory mechanism to recognize that infant cries are a signal of unhappiness, and thus be motivated to do something about it."[41]

Perhaps, as my grandmother used to assure me, infants sometimes cry because they are hungry, overtired, or (my favorite) "to exercise their little lungs." Nevertheless, various studies indicate that the major reason babies cry, at least in industrialized societies, is to reestablish physical contact with separated caregivers.[42] Newborns cry when separated from their mothers and then stop when reunited.[43] Experiments show that infants of up to four months of age cry when their mothers leave the room, which prompts the mothers to return and the infants to stop crying.[44] Even chimpanzee infants, who usually do not cry much, do so excessively during the first two months of their lives if they are separated from their mothers.[45] They, like tiny humans (except those who have colic), cease crying immediately when they are held. (Remember Joni's intense need to be held by Nadia Kohts.)

It is interesting to contemplate why crying peaks, on average, at approximately six weeks of age in both chimpanzees and humans, by which time an infant chimpanzee is developed enough to cling beneath its traveling mother without much help. And why do the cries of both species cluster in the late afternoon and early evening—the very time of day when chimpanzee mothers occupy all of their extremities by climbing high into the trees to build their night nests? Does crying peak when little chimps, at least, are especially vulnerable to falling off their climbing mothers? Are their cries equivalent to a human toddler's plaintive "carry me"—arms beseechingly held overhead?

Also, crying increases the strength of the grasping reflex in human infants.[46] (This was discovered in the 1930s during experiments that showed that the ability of young infants to support their weight by clinging with one hand decreased from monkeys to chimpanzees and was extremely limited in human infants despite their strong vestigial grasps.) Are the unique cries of tiny human beings compensating for the loss of little grasping hands and feet that once clung to ancestral mothers? Did the evolution of the unique form of human crying begin by maintaining physical contact between infants and their mothers? Perhaps. One thing seems clear: very often when a tiny human cries, he or she simply wants to be picked up and cuddled.

And what role does the human mother play in maintaining physical contact? The answer depends very much on whether Mom lives in an industrialized culture. Many segments of American society seek to develop independence in young infants and may, accordingly, permit them to cry for extended periods of time and to sleep separately from their parents, often in their own bedrooms. Feedings may be scheduled and are frequently delivered by bottle rather than breast. Western babies are also likely to be given solid foods and weaned much earlier than those living in traditional communities.

The care of infants is more physically challenging in traditional societies, where caregivers routinely breast-feed on demand for at least several years and maintain close physical contact with infants during the day. Mothers from these cultures also normally sleep with their infants for years, recognizing babies' need for round-the-clock nursing and cuddling.[47] Balinese infants are carried at all times during their first 210 days of life, while those in Turkish villages are not allowed to be alone or left to cry, and are picked up, held, and cuddled as much as possible. Ideally, Warlpiri infants are never denied the breast—a crying baby tells others that the mother is not doing her job well—and children sleep in their mothers' embrace until they are as old as eight. It is frowned upon if traditional Fulani mothers are separated from their infants for more than two hours a day before their babies are two years old. Ifaluk infants also are not supposed to be left alone. Not surprisingly, infants in such traditional cultures cry much less than those in industrialized societies.

But how do these mothers still accomplish day-to-day tasks while providing such consistent physical contact? The answer can be found in two innovations: one behavioral, one technological. Behaviorally, women have lost the chimpanzee mother's extreme reluctance to allow others to hold, carry, or care for their infants, a reluctance that may have developed as a mechanism for avoiding infanticide. We saw in Chapter 1 that chimpanzee youngsters are eager to play with and carry new siblings. As the infants mature, their mothers permit a limited amount of interaction with siblings but continue to keep a wary eye out. (Recall that Getty was ten months old before his mother allowed his grandmother to groom him.) Human children are also eager to interact with younger siblings, and their mothers are much more permissive in this respect than chimpanzee moms.

In societies where mothers work in the fields or gardens, such as the Beng and Ifaluk, babies are often cared for during the day by female relatives, other adults, older siblings (most often daughters), cousins, or other children. When the infants become hungry, their caretakers simply bring them to their mothers to nurse. These babysitters (technically known as allomothers) are commonly only a few years older than their charges. Whereas little girls in industrialized societies are frequently given dolls (like my nine-year-old granddaughter, Kylene, pictured on next page), those in other cultures (such as this little girl from Ethiopia) often have responsibility for the real thing.

Environmental conditions also largely determine how much mothers allow other children to care for their infants. For example, the Hadza make their living hunting and gathering in northern Tanzania.[48] Plenty of plant food is available within short distances, so mothers do not have to travel far to gather it. The environment is also safe enough that mothers feel comfortable leaving toddlers as young as two years old in camp, where they

FIGURE 2.2. *Left:* Ethiopian girl with her sibling. *Photo by Dean Falk.* *Right:* my nine-year-old granddaughter, Kylene, with her doll. *Photo by Michael Riddle.*

play relatively unsupervised with other kids of various ages. These children seem to get along and even do some food collecting on their own. The environment has enough resources and is safe enough that children can look after one another.

But it hasn't always been easy for hunters and gatherers to find suitable babysitters. Until recently, the !Kung San of southern Africa lived under very hot desert conditions where men engaged in hunting and women routinely went on long gathering expeditions, regularly carrying their two- or three-year-olds.[49] Again, environment played a pivotal role, because safe sources of drinking water were hard to come by, and mothers kept their infants hydrated with breast milk instead. Human breast milk, like that of chimpanzees, has a high water content, and San mothers nursed their infants up to four times an hour.[50] (The Beng, in fact, call breast milk "breastwater."[51]) Hrdy writes:

Even before the Neolithic, allomothers were employed *when safe allomaternal options were available.* It is beginning to look like the prevalence of exclusive and long-lasting mother-infant relationships might be an artifact of the harsh environments where hunter-gatherer lifestyles have persisted long enough for anthropologists to study them, an artifact of mothers needing to travel long distances to find food and water, of predators lurking nearby, and of settlement patterns where the mother's kin live far away.[52]

Ultimately, reliance on babysitters is a behavioral adaptation that evolved to support the long-lasting mother-infant relationships (similar to those of chimpanzees) that would have characterized our earliest ancestors.

The second innovation that allows women to remain in close physical contact with their helpless babies is technological: baby slings. Everywhere in the nonindustrialized world, babies are carried on their mothers' or babysitters' backs, hips, or bellies in baby slings. The Beng use cloth to tie babies onto the backs of girls as young as seven.[53] Traditional Balinese and Fulani mothers hold their babies in slings as they complete their daily tasks.[54] In the old days, Warlpiri mothers carried their babies along their sides in oblong bowls that were carved from wood and suspended over the shoulders with straps.[55] Some of the earliest baby slings probably were made of leather, which San women still use. Interestingly, both Beng and Fulani mothers give their infants enemas twice a day to avoid being soiled while carrying them in slings. According to Beng tradition, "This is good—then you can give your baby to a *leng kuli* (baby carrier) without worry that the baby carrier's clothes will be soiled as she carries the baby, for this would be a great shame on you!"[56]

Baby slings allowed mothers to not only carry their nursing infants, but also to gather and transport plant foods that became ripe in different locations. This carry-all aspect of slings would have been key to the shift from chimpanzee-like foraging (eating on the go) to the gathering of various foods and transporting them back to a home base. We'll see later that the emergence of such

"Central Place Provisioning," as Marlowe has dubbed it, may have been crucial for the eventual emergence of human social values.

MOTHER-INFANT COMMUNICATION: A MUSICAL DIFFERENCE

Human mothers and babies communicate using much of the same body language as chimpanzees. Three-month-old humans intently watch their mothers' faces and express themselves with their own little faces and bodies. Mother-infant playing, tickling, and touching are much like that of chimpanzees. Laughter occurs during the play of both humans and apes, although humans produce it differently, which has important implications for understanding the evolution of the breath control needed for speech.[57] Mothers of both species also examine, cradle, pat, nuzzle, and kiss their infants.

By the time human infants are three to four months old, they have intense emotional relationships with their mothers expressed through mutually coordinated attention, movements, and facial expressions. Mothers all over the world make exaggerated friendly faces at their babies, which have precursors in other primates. These expressions include raising and flashing the eyebrows, bobbing the head, smiling, and nodding. Mothers and their babies are sensitive to each other's behaviors and each "is able to enter the temporal world and feeling state of the other."[58]

Despite the similarities between chimp and human mothers, the human species differs significantly from chimpanzees when it comes to mother-infant vocal communication. As we have seen, human infants have evolved distinctive ways of crying to get caregivers' attention while communicating their own needs and moods. Our infants cry to get what they want. In turn, mothers soothe distraught infants by making special sounds that probably evolved long before the use of baby slings, and perhaps even before the regular use of babysitters. I am not referring to baby talk, but rather to another type of vocal communication that may have predated it—lullabies.

Lullabies are used throughout the world to soothe infants and lull them to sleep.[59] They are distinctive from other songs—so much so, in fact, that adults listening to songs in unfamiliar languages can distinguish the lullabies from the equally slow non-lullabies.[60] And babies love them. Experiments have shown that infants prefer listening to lullabies over adult songs, especially if they are sung by women. Other caregivers, however, including men and children, also sing lullabies to varying degrees, depending on the culture, and, like moms, they know intuitively how to suit the needs and listening capabilities of their young audiences. In addition, mothers generally sing lullabies in a more expressive manner than fathers, while dads are more expressive with infant sons than daughters.

Mothers around the world don't just sing their infants into slumber, they also rock them:

> The rocking, jostling, and lifting of infants, which mothers enjoy so much, accommodates the infantile need for vestibular stimulation. An excited infant can be calmed by rocking it. In kin-based societies infants spend most of their time being carried about by the mother or another person. Vestibular stimuli communicate to the infant that he is not alone. Hospitalized children isolated from this stimulus often develop movement stereotypes, like rocking and self-patting, that serves as self-stimulation.[61]

One familiar Western lullaby underscores the relationship between singing and rocking infants, and it refers to a fall from the trees, evoking the image of a chimp infant dangling precariously. One might therefore suggest (somewhat whimsically) that this lullaby soothes a primordial fear retained from when our ancestors still slept in tree nests (cradles of boughs), as chimpanzees do. (Primary stresses are capitalized and underlined; secondary stresses are underlined):

ROCK-a- // bye / ba- // by / ON the // tree / top ///

WHEN the // wind / blows // the / CRA- / dle / will / rock ///

WHEN the // bough / breaks // the CRA- / dle /will / fall ///

And /// DOWN / will / come / Ba- // by / CRA- / dle / and / all /// [62]

But what do lullabies convey to babies? Do they understand the messages? The rhythm, regularity, and simple structures of infant-directed songs help to shape and control their emotions.[63] Mothers sing at fast tempos (as in many play songs) to attract their infant's attention, and they sing at slower tempos to maintain that attention. Infants participate in coordinating the structure of these interactions with their emotional reactions, such as smiling, cooing, and jerking their limbs. The song's meaning manifests itself in the emotions it evokes in the infant.

Babies do understand. Even without words, infants grasp the emotional content of songs. This seems to be universal in humans, like the ability of adults everywhere to understand the emotional nuances of vocalizations.[64]

Next we'll see how the distinctive cries of human babies and special vocalizations from their mothers both began to evolve when our ancestors started walking on two legs and their infants lost the ability to cling to their mothers. Before the invention of baby slings, when little hands and feet could no longer hang on, and moms sometimes could not cradle their babies, these vocalizations would have been the best way to sustain mother-infant bonds.

Chapter 3

AND DOWN WILL COME BABY

Hush-a-Bye

Rock-a-bye, baby, on the treetop!

When the wind blows the cradle will rock;

When the bough breaks the cradle will fall;

And down will come baby, cradle and all.

—*Mother Goose*

A CHIMPANZEE BABY LEARNS the vicissitudes and pleasures of life while clinging to its mother's chest and, later, from the vantage point of her back. At first the mother helps to keep her newborn attached, but by about two months of age, the infant is developed enough to hold on to her underside without assistance. There, it suckles at leisure as Mom knuckle-walks on the ground or swings through the trees. Infants' ability to travel attached to their mothers is vital for nonhuman primates, as illustrated by a mother-infant pair Jane Goodall studied. The mother, Madam Bee, had successfully raised two infants before becoming partially paralyzed from polio. When traveling, Madam Bee was unable to physically support her most recent newborn, Bee-hind, who responded by whimpering, screaming, and repeatedly falling off her mother. Bee-hind died when she was only a few months old, leaving a corpse full of wounds and scratches.[1]

Physical contact is equally important to human babies. To glean more information about the contact between human mothers and their babies, I relied on several young informants. (When studying unfamiliar people and places, anthropologists often consult people from the culture in question.) One of my informants, Sisa Uzendoski, was born on July 23, 2003, and is the daughter of Michael, a cultural anthropologist and colleague at Florida State University, and Edith, a full-time mom who grew up in the Amazonian region of Ecuador. I learned that although Sisa's crib is next to her parents' bed, she spends much of the night snuggled in with them. Sisa has never used pacifiers or taken bottles. When her parents take her to visit family in Ecuador, she is cared for by twenty-three first cousins. Sisa has also learned three languages simultaneously: English, Spanish, and Quichua.

Another informant, Josie Parkinson, was born about three months after Sisa. Josie is Chinese, and her American parents, William and Betsy Parkinson, adopted her in November 2004. Betsy teaches first grade and Bill is an archaeologist. Josie did not learn to crawl because she was never placed on the floor in the orphanage where she spent her first year. After she was adopted, however, Josie was routinely placed on the floor, where she invented an amazingly rapid and efficient way of scooting along on her bottom (she could carry things in her hands while moving). Because Betsy uses American Sign Language (ASL) in her teaching, Josie has learned both English and ASL.

During the course of hominin evolution, human babies lost the ability to ride on their mothers.[2] Nevertheless, our infants still have a deeply rooted longing for close physical contact with caregivers. Josie began to talk by the time she was nineteen months old. One of her first words was "up," spoken directly in front of her parents and accompanied by an arms-out, "pick-me-up" gesture. She liked to cuddle with her special blanket, and she loved to be rocked as her father sang a lullaby at bedtime. Sisa never had a special blanket or snuggle toy, and neither did four-year-old

FIGURE 3.1. Josie Parkinson *(left)* and Sisa Uzendoski *(right)* discuss the finer points of learning to read with a friend, Grant Gravlee.

Claire, another daughter of friends. However, Sisa continued to nurse often, and Claire still called to her parents in the night, when they would bring her to their bed and allow her to sleep snuggled against her mother's back. Although these children did not spend the day attached to their mothers' bodies the way chimpanzee infants do, their need for snuggling and intimate contact with their parents seemed just as strong as the ill-fated Beehind's. Primates' need for sustained contact between mothers and infants is so strong that it even appears in our more distant relatives, monkeys. This suggests that the infant ability to cling is deeply rooted in our ancestry, making its loss in human babies even more significant.

THE HARLOW EXPERIMENTS

In famous experiments psychologist Harry Harlow carried out on monkeys in the 1950s and 1960s, newborn macaques were

FIGURE 3.2. A baby macaque clings to
a cloth-covered surrogate mother. *Courtesy
of the Harlow Primate Laboratory, University of
Wisconsin-Madison.*

isolated in cages that were cushioned with soft cloths. When the
cloths were removed, the infants threw temper tantrums similar to
those of human babies who had become attached to special
"blankies" or stuffed toys. This simple observation sparked a series
of experiments where newborn monkeys were taken from their
mothers and raised with two artificial surrogate "mothers." One
mother was constructed of wire mesh and had a single nipple in
the center of its chest that provided nourishment. The other
lacked the formula-giving nipple, and its frame was padded with
soft, tan terry cloth. "The result," noted Harlow, "was a mother, soft,
warm, and tender, a mother with infinite patience, a mother available twenty-four hours a day, a mother that never scolded her
infant and never struck or bit her baby in anger."[3] Although monkeys and apes in the wild cling to their mothers constantly and
nurse regularly, Harlow discovered that the isolated macaques

spent the vast majority of their time clinging to the nonlactating, cloth-covered surrogate and visited the wire mother only to satisfy their hunger.

As the infants grew, their preference for the cloth mother increased. If the cloth mother was removed from the cage, the infants crouched and frantically clutched their bodies. When an infant and the cloth surrogate were put into a strange environment containing curiosity-stimulating objects, the infant initially rushed to and clutched the surrogate, which in time became a source of security. Harlow also observed that newborns were at first inattentive to the surrogate's face, but by the time they were a month old, the simple wooden ball with bicycle reflectors for eyes had become an object of special attention. In one experiment, the surrogate's face was replaced with a more realistic monkey mask, which caused the infant to scream, rush to a corner, and violently rock, presumably because it was no longer "Mom's" face. The cloth surrogate had become the only mother these monkeys would ever know. Unsurprisingly, when some of the monkeys matured and gave birth, they were utterly inadequate as mothers.

Harlow's experiments showed not only that infant primates have an intense need for intimate physical contact with their mothers, but also that such "contact comfort" is crucial to develop a capacity for love and normal emotional growth. He even suggested that the main function of nursing was to facilitate infants' emotional development by ensuring frequent contact with their mothers. His conclusion—that newborn primates' need for cuddling is as ingrained as the desire for food—was radical for the 1950s. Although his experiments would not be allowed today, Harlow is credited with helping to reverse the then-fashionable American trend of refusing to coddle babies. This philosophy had stemmed from the child-rearing advice of psychologist John Watson, who established the American school of behaviorism: "Never hug and kiss them, never let them sit on your lap. If you must, kiss them once on the forehead when they say good night. Shake hands with them in the morning."[4]

Harlow's findings can also apply to human infants. Although human offspring can't cling autonomously to their mothers, they continue to be born with strong vestigial grasping reflexes. Like Harlow's monkeys, human babies instinctively seek intimate contact with caregivers. Research on American infants suggests that the main reason babies cry is to reestablish physical contact with separated caregivers.[5] Consistent with this, crying increases the strength of infants' grasping reflexes.[6] Harlow's suggestion that prolonged nursing is more about contact comfort than nutrition also rings true for human infants in light of the calming effects of pacifiers and thumbs.

I believe that the blankets our children cling to are stand-ins for the fuzzy chests of ancestral mothers. However, the monkeys' attachment to soft cloths resulted from artificial conditions of social isolation. Because of natural evolutionary changes, human babies, on the other hand, have lost the ancestral ability to cling. In a sense, we can think of our babies' helplessness as the result of an evolutionary "natural experiment" that severed prolonged physical contact between hominin mothers and infants.[7] Early mothers, however, were not completely removed from their infants, as they were in Harlow's experiments, and this made a significant evolutionary difference. But before we describe the critical prehistoric experiment, we will review evolution itself, and how it operates on mothers in particular.

How Evolution Works

Although evolutionary biologists quibble, they wholeheartedly embrace the theory of natural selection Charles Darwin and Alfred Russel Wallace developed in the mid-nineteenth century. In fact, so much evidence has been amassed in favor of the theory that, like gravity, it may be considered a law. The concept of natural selection is easy to grasp and elegant in its logic: Species give birth to more offspring than available resources can support. Those individuals with anatomies and behaviors that make them better at competing for limited resources survive and reproduce, and their genes shape future populations.

The competition for resources is usually subtle and unconscious. Adaptive traits that favor survival might include a successful immune response to certain diseases (such as the plague or AIDS), a physical appearance that blends in with the environment, or a tendency toward harmonious relationships. Because such characteristics frequently are genetically based, offspring may inherit them. Through this simple mechanism, adaptive features are naturally passed to future generations, while unfavorable ones are not. If this process continues long enough, natural selection may change an isolated population until its members can no longer interbreed with those from the populations that it once resembled, and a new species is born.

The key to natural selection is successful reproduction. To be "selected for," however, it is not enough for an animal to have offspring. The offspring themselves must be able to breed successfully. For mammals that nurse their young, this means mothers are the primary, although not exclusive, gatekeepers for their species' genetic future. If an infant is raised to sexual maturity, copies of its parents' genes are likely to find their way into the next generation.[8]

Natural selection favors any behavior that allows an individual or group to out-reproduce others. These behaviors are called reproductive strategies, though animals (including humans) are unaware that their behaviors are strategic in an evolutionary sense. For example, an older man's choice of a younger woman may be viewed as a reproductive strategy since her fertility enhances his chances of having offspring. The older man would probably be taken aback, though, at the suggestion that the underlying reason for his choice was that his male ancestors preferred fertile women and were differentially selected for. All too often, primatologists and anthropologists write as if animals (human or otherwise) consciously worry about how to maximize their reproductive chances. However, these impulses are almost always unconscious.

Because humans are primates, monkeys and apes provide excellent models for exploring past and present human behaviors. Like many animals, male and female primates usually differ in their reproductive strategies. Since males do not become pregnant and therefore do not require time to bear and raise young, there is no limit to the number of offspring they can produce, and it is in their evolutionary interest to impregnate as many females as possible. Depending on the species, behaviors that increase a male's mating opportunities may include befriending and defending females, helping them carry or feed their infants, and achieving high status within the male community.

In primates such as gorillas, howler monkeys, and some kinds of baboons, one male frequently lives with many adult females in so-called harems. In these one-male groups, males engage in physical combat with outsiders who attempt to drive them from their groups. Compared to females, adult males are much larger and have larger teeth—anatomical weaponry that has been selected for in conjunction with the male reproductive strategy of direct male-to-male combat. Another male reproductive strategy involves infanticide during attempts to take over one-male groups, where an invading male kills nursing infants, causing their mothers to resume ovulation and become sexually active. The new resident male then can impregnate the group's females, thereby increasing his reproductive fitness at the previous leader's expense.

Other primates, such as chimpanzees and woolly spider monkeys, live in multi-male, multi-female groups. In these groups, males are somewhat larger than females, but not to the degree seen in the one-male groups. Sexually receptive females in multi-male groups mate with numerous partners. Although the males do not fight over females as much as males living in one-male groups do, they have enlarged testes, which are thought to facilitate competition between the sperm of potential fathers *after* mating, through a mechanism called sperm competition.

The reproductive strategies of monogamous males, such as gibbons, on the other hand, also differ from the direct-combat model in one-male groups. In these species, the bodies and teeth of the two sexes are about the same size, and males usually compete with each other in vocal contests.

Since females can give birth a limited number of times, it is in their evolutionary interest to conceive offspring that are healthy enough to grow up and reproduce. Females select mates who show signs of having good genes, an unconscious reproductive strategy known as female choice. A sexually receptive female primate might prefer males of high status, big males, or males with impressive coats of fur. Females might also mate preferentially with males who share food, defend them, or play with their infants. Such males are likely to help with future offspring. Women exhibit similar preferences but would probably attribute their desires to sexual attractiveness rather than a quest for positive paternal genetic contributions.

Because bearing, nursing, transporting, and raising young are physically taxing, mothers must obtain the necessary resources such as food, water, and shelter. Rather than fighting over the opposite sex, females are more likely to compete for resources. One strategy is to achieve a high rank within the female hierarchy. Although female social hierarchies are not present in all species, a female's success at obtaining food and producing offspring correlates significantly with her rank in some hierarchical species.

Nonhuman primates do not understand the concept of fatherhood, and the responsibility for raising infants falls heavily on mothers. Adult females living in one-male groups must protect themselves and their offspring from aggressive males, so they defend their babies against potential attackers in groups. Similar "aunting" behavior is common in various kinds of social groups and in other aspects of infant care. Such behavior is also more likely to occur among related females. Although relatives such as aunts and grandmothers can be vital to an infant's survival, contemporary evolutionary theory (supported by Harlow's findings) suggests that the most important individual in a newborn's life is the mother

When some of Harlow's monkeys were artificially inseminated and gave birth, the new mothers proved to be so incompetent that their infants never would have survived in the wild. Because they had been isolated and deprived of physical contact with their own mothers (or any other monkeys, for that matter), these monkeys were selected against, albeit unnaturally. During hominin prehistory, our ancestors' infants were also deprived of constant physical contact with their mothers, but under completely natural conditions. The survivors of this natural experiment fared much better than Harlow's monkeys. In fact, they set the stage for what our species would become.

BIPEDALISM: CATALYST FOR PREHISTORY'S EXPERIMENT

When hominin and chimpanzee lineages diverged from a common African population 5 million to 7 million years ago, our ancestors started down a path that would lead to one of the most critical natural experiments in human prehistory.[9] In this natural experiment, as in Harlow's, infants were deprived of the constant warmth and protection of their mothers' bodies as our ancestors shifted from walking primarily on four legs (quadrupedalism) to walking on two (bipedalism).

Although many paleoanthropologists think that selection for bipedalism was *the* crucial factor that caused hominins to diverge from apes, they are still unsure about why upright walking became a target of natural selection. Was its primary function to free the forelimbs to make better stone tools or to carry food, water, branches, or rocks? Or did bipedalism begin when hominins began standing up to gather fruits, nuts, and insects from trees, or to survey the surrounding terrain for game or predators by peering over tall grass? Since the ability to run marathons is unique to people, could the primary advantage of bipedalism have been the ability to outrun game (persistence hunting)?[10] Or perhaps the initial advantage was to minimize the amount of skin exposed to damaging ultraviolet rays from the strong African sun.[11]

These hypotheses are impossible to disentangle, though some have fallen out of favor because of newly discovered fossils that may be hominins. In the summer of 2002, a French team led by Michel Brunet unearthed the earliest-known candidate for a hominin. They discovered the fossil in Chad, in central Africa. This new species, *Sahelanthropus tchadensis* (nicknamed Toumai, meaning "hope of life"), surprised paleoanthropologists because of its age (perhaps 7 million years old) and because it had lived in a lush forest rather than the expected savanna grasslands. This presumed biped would not have stood up to peer over tall grass, nor would sunburn have been a problem. It's also unlikely that Toumai used its freed forelimbs to make stone tools, since none appear in the fossil record for at least 3 million more years. The same observations apply to a slightly younger (about 6 million years old) potential hominin, *Orrorin tugenensis* ("Original Man from Tugen"), which Brigitte Senut and Martin Pickford discovered in Kenya in 2000. Scientists seem to still be far from discovering why our ancestors adopted a vertical lifestyle.

Although there is no consensus about why bipedalism developed, the fossil record unequivocally shows that it had appeared in an important group of early ancestors, known as australopithecines. But the shift to bipedalism was not rapid. One of the most complete fossils providing relevant information is the famous 3.2 million–year-old partial skeleton from Ethiopia, Lucy. Though she was an adult, Lucy stood only three and a half feet tall. Her pelvis and leg bones show that she was bipedal, but Lucy also had long apelike arms and curved finger bones, suggesting that her species, *Australopithecus afarensis,* spent time—and probably even slept—in the treetops. The arrangement of muscle attachments on Lucy's pelvis indicates that her bipedal gait was less fluid than that of humans. Chimpanzees waddle when they walk on two legs, and Lucy probably waddled a bit, too.

However, Lucy's pelvis was not exactly like a chimpanzee's. Instead, it was shortened and curved slightly forward on the sides, somewhat like ours. The ball that connected the top part of Lucy's thigh bone to her pelvis had a long "neck," indicating that the evolutionary broadening of females' hips had not yet occurred in her species. And the adult and infant brain sizes of early australopithecines were in the chimpanzee range. This indicates that although their pelvises clearly had become modified for bipedalism, the females of Lucy's species continued to give birth relatively easily, as living apes do.[12] The prolonged and painful childbirths women experience would come later, when a combination of bigger brains and narrower pelvises created an unbearably tight fit between babies' heads and mothers' birth canals. Only when babies began to be born increasingly immature and helpless did the natural experiment of mother-infant separation truly begin.

Traces of small, long-armed creatures like Lucy appeared for millions of years in the fossil record. For example, the partial skeleton OH 62 from Olduvai Gorge in Tanzania resembles Lucy and dates to about 1.8 million years ago (nearly a million and a half years more recent than Lucy). In what may be the most controversial hominin discovery in more than eighty years, an approximately three and a half foot tall adult woman with an australopithecine-like pelvis and limb proportions was discovered in Indonesia in 2004.[13] Many believe that this specimen, nicknamed Hobbit, represents an entirely new species of human (*Homo floresiensis*) that lived until a mere seventeen thousand years ago on the island of Flores in Indonesia. This discovery was shocking considering the long-held conviction that *Homo sapiens* was the only hominin that existed so recently.

No one can be sure of the relationship, if any, of these fossils to each other, but they all show that primitive, small-bodied hominins existed for a long time. That australopithecines also had small ape-sized brains and pelvises that were not quite modern shows that giving birth had not yet become traumatic. For our earliest ancestors, there was no need for natural selection to favor newborns who were less developed and helpless. That would happen later, when babies' heads became too large to navigate constricted birth canals. Prenatal australopithecines still had plenty of time to grow in the womb and, like monkeys and apes, were born developed enough to cling to Mom, at least with their hands.[14] We will see in the next chapter that their feet may already have become so modified for walking that they lost some of their gripping power.

Another telling fossil from East Africa, KNM-WT 15000, dates to 1.6 million years ago. Discovered by Alan Walker and Kamoya Kimeu, this specimen hails from a site called Nariokotome in West Turkana, Kenya, and is the earliest relatively complete skeleton known for *Homo erectus*. (It also wins the prize for having the most nicknames of any hominin fossil: "15K," "Nariokotome," "Turkana lad," and "the Strapping Youth.") When its

discovery was announced in 1985, this now-famous skeleton star-
tled paleoanthropologists because its body differed so radically
from those of its australopithecine neighbors. WT 15000 is the
fossilized remains of an eight-to-eleven-year-old boy whose arms
and legs were proportioned like those of modern people. This
youth had already grown to about five foot five and probably
would have been over six feet tall as an adult—not at all resembling
short, long-armed Lucy or 15K's near contemporary, OH 62.[15]

Although the strapping youth was tall and proportioned like
modern humans, his pelvis was much narrower—so much so that
some surmise that the pelvises and birth canals of *Homo erectus* females were also very narrow and that their babies were probably
born smaller than ours,[16] although this is controversial.[17] Anthropologists traditionally measure the volume of the braincase,
known as cranial capacity, to estimate brain size. WT 15000's cranial capacity was just under 900 cubic centimeters (cm^3) and
would have been only slightly larger had he lived to adulthood.[18]
This estimate is twice the average for adult australopithecines
(450 cm^3) and two-thirds that for modern humans (1350 cm^3).

Childbirth was probably becoming difficult for *Homo erectus*
women compared to their australopithecine ancestors because of
their evolving pelvises and increasing brain size. If so, *Homo erectus*
newborns would have undergone a significant amount of brain
development during their first year, but perhaps not to the degree
of our own babies. Although the partial postponement of brain
growth until after birth helped to accommodate changed pelvises,
it also came at a cost. Delayed neurological development in *Homo
erectus* babies would have corresponded with slower motor development. Clinging autonomously would have been increasingly
difficult.

A *Homo erectus* mother likely carried her helpless baby while
collecting food, water, and other resources by using a baby sling.
Unfortunately, we do not know when baby slings first appeared.
However, as noted by Adrienne Zihlman of the University of
California at Santa Cruz, they were probably among the earliest
tools invented.[19] Because textiles do not appear in the archaeological record until relatively recently,[20] *Homo erectus*'s baby slings

FIGURE 3.3. A woman carries a baby in a sling in
Ethiopia. *Photo by Dean Falk.*

probably were not made of woven vegetation, as some anthropologists have suggested. Instead, they may have been made from the
hides of the animals that were scavenged or hunted.

What happened before the advent of baby slings? To answer
that question, we must consider the ancestors of *Homo erectus*,
who split from australopithecines at an unknown point before 2
million years ago. At that time, bipedalism began to collide with
increasing brain size. As evolution began transforming small, ape-
proportioned bipeds into tall, gracefully striding humans, such as
the *Homo erectus* boy from Turkana, our ancestors' babies would
have had more trouble remaining attached to their mothers' bodies. So the natural experiment indicated by 15K's pelvis and head
may have started well before 1.6 million years ago.[21]

BEFORE BABY SLINGS

Consider the severe evolutionary pressures for the ancestors of *Homo erectus*. As the well-documented trend toward enlarged brains continued, mortality increased in full-term babies whose heads were too large to fit through increasingly narrow birth canals. Their genes were eliminated from the gene pool not just through their own deaths, but also through the frequent deaths of their mothers during childbirth. Many larger babies and their laboring mothers would have died during the long and difficult transition to full-fledged bipedalism. Natural selection must have been so severe that the increasingly narrowed hominin birth canal can be seen as a kind of evolutionary bottleneck that drastically reshaped our species. However, developmental timing of gestating infants would have varied. The babies whose heads remained relatively small until after they were born successfully negotiated tricky birth canals and survived. The dramatic shift to helpless, immature newborns began when natural selection started to favor slow developers.

This process would have been gradual and incremental. At first, birthing would simply have become longer and more painful in the ancestors of *Homo erectus*. As natural selection continued to nudge brain sizes larger, however, mother/infant mortality would have emerged at unprecedented levels and, over time, the surviving newborns would have postponed some brain development until after birth. Compared with their chimpanzee relatives, these developmentally premature babies would have depended more and more on their mothers for physical support.

Chimpanzee mothers initially use an arm to support their babies, but this doesn't last long because infants soon are strong enough to cling independently with their hands and feet. Early hominin mothers undoubtedly kept the chimpanzee habit of cradling newborns in their arms. But as brain size continued to evolve, infants' ability to ride unaided on their mothers would have become increasingly delayed until they eventually failed to develop the necessary motor coordination and strength. These infants soon would grow too heavy for Mom to carry. What would she have done?

FIGURE 3.4. Michaelmas, the one-year-old son of Miff, clings to his mother using both hands and feet. *Photo by Curt Busse.*

She could have cajoled others to share the task of carrying her infant. Adolescent chimpanzee females are intensely interested in babies, and this was likely also the case for early hominins, so aunting behavior probably increased to some degree in the female ancestors of *Homo erectus*. However, such aunts would soon bear and carry their own infants, and they would also need to forage for food. Adult males would have been even less helpful because, if Goodall's chimps are any indication, they would have spent much of their time away from females and youngsters. The natural experiment began long before signs of settlements appear in the archaeological record, so mothers did not leave their infants in others' care at a home base. Some mothers may have stashed their infants in temporary hiding places, such as dense foliage, but predators would have made this a poor solution.

I believe that the lactating ancestors of *Homo erectus* continued to carry babies in their arms and on their hips when they looked for food, water, or shelter. (Perhaps trekking hominin mothers unconsciously supported their young with their left arms because babies preferred to be close to the comforting maternal heartbeat. If so, could this habit have contributed to the development of right-handedness in people?) Traveling mothers would have stopped periodically to relax with their babies, as ape mothers do. When foraging, however, mothers would have needed both hands to pick berries, dig for roots, or gather other resources, and "aunts" or siblings would not always have been available to hold their infants. Only one option would have remained: to put the baby down. Early moms would have remained very close to their separated infants, of course, and kept a wary eye out even as they worked. For the first time in prehistory, babies were routinely deprived of their deeply ingrained need for constant physical contact with their mothers.

Like poor little Bee-hind (not to mention Harlow's monkeys), the infants must have been extremely distressed by this development. In fact, as illustrated by my young informants, modern babies clearly still prefer to be in constant physical contact with their parents. When I began writing this book, Josie had just discovered how to throw a temper tantrum. She had these tantrums on the uncarpeted floor, where she could beat the palm of her hand loudly. Shortly after learning to walk, Josie began throwing tantrums while standing up and stomping on the floor. I asked Josie's mom, "Why is she having these tantrums?" "Because she wants to be picked up," she answered, looking surprised that I even had to ask.

Although it was undoubtedly traumatic for our ancestors and remains difficult for babies like Josie today, I believe the loss of constant contact between mothers and infants was pivotal for humankind. Most important, it helped our ancestors find their voice.

Chapter 4

HOW OUR ANCESTORS FOUND THEIR VOICE

Do Not Cry

Do not cry, my child,

O do not cry, my little one,

For I, your mother, am here.

O do not cry, my child.

—Nigerian lullaby

Long before baby slings were invented, mothers put their helpless babies down nearby as they went about their daily tasks. Our ancestors' infants probably would not have enjoyed being separated, even temporarily, from their moms any more than Harlow's monkeys or chimpanzee infants did. And like our own babies today, babies of our early ancestors would have whimpered and cried to protest being removed from their mothers.

I believe that at this point in evolution, mothers started maintaining *vocal* contact with their infants. Soothing voices would have periodically substituted for cradling arms as busy mothers lulled their babies to sleep or reassured them that they were near. I have dubbed this idea the "putting the baby down" (PTBD) hypothesis.[1] It is consistent with new paleoanthropological discoveries from the gap between seven million and five million years ago, when our ancestors branched off from chimpanzees, and 1.6 million years ago, by which time *Homo erectus* probably had both baby slings and a primitive form of language. It was during this long, mysterious transition that our ancestors began to communicate verbally.

THE NIGHT SHIFT

Hominins underwent key lifestyle changes that sparked the development of language, and the move from trees to the ground was particularly pivotal. We once thought the earliest hominins perfected bipedal travel in open grasslands, but the discovery of fossils from ancestors who lived about seven million to three million years ago in wooded parts of Africa has disproved that theory. Fossils indicate that *Sahelanthropus, Orrorin,* and Lucy *(Australopithecus afarensis)* all seem to have walked on two legs. A few hominin finger and toe bones have also been found in wooded African areas, and those bones turned out to be a bit curved, like those of tree-dwelling apes, rather than straighter, like our own. Lucy's relatively long arms added to the suspicion that her kind spent time in trees as well as on the ground. Martin Pickford of the College of France, Paris, says, "It is considered most likely that bipedalism evolved in forested to well-wooded environments, and only later did bipedal hominids venture into more open country. Thus bipedalism was a development which subsequently enabled hominids to invade open country. It was not invasion of open country that led to bipedalism as so often thought."[2]

Clearly, our ancestors not only continued to live in wooded or forested habitats for quite some time after they split from chimpanzees, but as they were perfecting their newfound bipedalism, they also still spent time in trees—for good reason, of course. Trees offer an escape from hungry leopards or lions, and they provide a secure haven for sleeping. Our earliest ancestors would

FIGURE 4.1. Chimpanzee Mike builds a safe sleeping nest in a tree at Gombe. *Photo by Curt Busse.*

have inherited the chimpanzee-like habit of climbing into trees at night to build comfortable sleeping nests, and it would be surprising if they had not passed along such a safe behavior to future generations.

Monkeys often sleep in trees or on cliffs to avoid predators, but they do not construct nests. The special behavior of climbing a tree, finding a convenient fork, and bending and weaving branches into a sleeping nest evolved only in the larger-bodied great apes to help prevent them from falling and to provide camouflage. Once the nest is prepared, the ape climbs in (with her infant if she has one) and settles down to sleep, often after calling back and forth to other apes nesting nearby. Like people, great apes sleep horizontally. And like great apes, people still sleep in nests (we call them beds)—though we tend to reuse the same nest night after night.

Eventually, of course, our ancestors stopped sleeping in trees. Most likely, this change occurred because fickle African climates caused the regions that apes and hominins inhabited to become cooler and drier, and trees began to disappear. By three million years ago, the weather had become noticeably more seasonal and arid, and during the next two million years, open grasslands gradually replaced wooded areas.[3] As the environment changed, the ancestors of chimpanzees and gorillas became confined within the few remaining wooded areas, or refuges. Descendants of these primates still live in some of the same African wooded areas today. Our ancestors, however, took a different tack. When the trees dwindled, they started making their nests on the ground, as the largest gorillas do today.[4]

This had profound implications for human evolution. Psychologist Frederick Coolidge and archaeologist Thomas Wynn believe our ancestors' sleep patterns changed, distinguishing humans from chimpanzees in response to the increased vulnerability from nocturnal predators associated with sleeping on the ground.[5] Early hominins probably decreased the total time they spent in bed while increasing the amount of rapid eye movement (REM) sleep, from which they could be awakened more easily.[6] REM sleep is important for modern humans because it helps us consolidate the patterns entailed in learning, for example, grammar or how to play chess.[7] This probably explains why our newborns spend about four times as long as adults in REM sleep, often accompanied by sucking movements and quivering eyelids. REM sleep is also a time for vivid dreams, simulating threat situations, rehearsing social interactions, and making subtle connections that lead to flashes of insight or inspiration.[8] Coolidge and Wynn suggest that the shift to ground sleeping sparked new sleep patterns that, over time, contributed to the evolution of creativity and innovation. (We'll learn more about this in Chapter 9.)

Despite this benefit, these must have been difficult nights for our ancestors. The dangers of ground sleeping became even more treacherous after three million years ago as babies were born increasingly helpless. Many tots would have fallen off their mothers and been mortally wounded during this arduous transition. The fossil record suggests that our ancestors had adopted a fully terrestrial way of life that most likely entailed the use of baby slings by the time of the strapping youth, 1.6 million years ago. But how long before this had mothers started putting their babies down and murmuring to soothe and calm them? When did motherese first come into play?

THE PITTER-PATTER OF LITTLE FOSSIL FEET

In 2006, the discovery of a 3.3-million-year-old skeleton of a three-year-old girl was announced.[9] The remains came from a region in Ethiopia called Dikika, so this toddler is known affectionately as the Dikika baby.

Lucy and the Dikika baby are from the same species, *Australopithecus afarensis,* and their remains were found just six miles apart. Unlike Lucy, however, the Dikika baby not only has a face and a complete torso, including the shoulder blades, but also a small throat bone called the hyoid bone, which has been central in discussions about language origins and had never before been found in australopithecines.[10] Most important, the baby has several intact finger bones and a nearly complete foot, both of which provide vital information about australopithecines' lifestyles.

Some details of the Dikika baby's anatomy, such as the size of her brain and the shape of her inner ear, nose, front baby teeth, and hyoid bone, resemble those of juvenile apes. Her shoulder blades face upward where the arm attaches, similar to a gorilla's, rather than facing to the side, like a human's. The orientation of the shoulder blades determines the carriage of the arms and, to some extent, arm movement. Gorillas' shoulder-blade orientation is consistent with their frequent extension of the arms overhead during activities such as tree climbing. The Dikika baby's shoulder-blade orientation indicates that others of her species did the same, especially since they lived in woodlands. Her finger bones are also consistent with this because, unlike yours and mine, they are long and curved. In fact, "one finger was still curled in a tiny grasp."[11] It seems that this baby could still hang on.

The lower part of her body, however, may suggest otherwise. Like those of the others of her species, the Dikika baby's legs were built for walking, as shown by the shape and angles of her thighbones, kneecaps, lower leg bones, and the heel and sole of her left foot. Her remarkable skeleton reveals that the legs and feet were evolving ahead of the rest of the body and that her species was moving around both on the ground and in trees. As Zeresenay Alemseged notes, "I see *A. afarensis* as foraging bipeds . . . climbing trees when necessary, especially when they were little."[12]

But was the Dikika baby still capable of a good chimpanzee-like grip when it came to hanging onto her mother? Although her grasping little fingers would suggest so, her feet raise some doubt. Apes have flexible big toes that stick out from the sides of their feet, much as our opposable thumbs do. Human big toes are more rigid and aligned with the other toes, which is a more efficient design for distributing the body's weight during walking. When we shifted to bipedalism, our feet ceased to function as a second pair of hands, our big toes became non-opposable, and we lost the ability to cling with our feet.

Although the Dikika baby's left heel and sole are consistent with bipedalism, we do not know whether her foot retained a grasping big toe. A separate fossilized big toe that belongs to the Dikika baby is encrusted in sandstone. When it is extracted, we will learn whether the baby clung to her mother with both hands and feet or struggled to stay attached using only her fingers.

Hominin mothers would have had to expend more energy supporting and carrying babies who could not hang on with both hands and feet. So if Dikika had a nongrasping big toe, mothers may have begun putting down their babies even before three million years ago. This would have been only the first step that led to a drastic change in prehistoric child care, however. After the Dikika baby died, wooded areas decreased significantly, hominins began to sleep on the ground, and their combination of enlarged brains and constricted birth canals caused babies to be born smaller, less developed, and even more dependent on their mothers. By then, vocal communication between mothers and nearby infants would have become increasingly important.

Other tantalizing hints in the fossil record suggest that the full transition to ground sleeping took a surprisingly long time. For most of prehistory, our ancestors lived in Africa. Around two

FIGURE 4.2. The Dikika baby is held by her mother. Note that the baby is portrayed with a non-grasping big toe, although this is not yet verified. *Image courtesy of Chris Sloan, National Geographic Society.*[13]

million years ago, some began migrating out of Africa, beginning the long process that led to the human colonization of most of the planet. These first migrations were on foot, and paleoanthropologists long assumed that the first wanderers looked something like the tall, human-proportioned 1.6-million-year-old boy, WT 15000.

However, recent discoveries show that hominins who looked more primitive had left Africa and settled in Eurasia by around 1.8 million years ago. The skulls of these hominins, which were found in Dmanisi, Georgia, have features that are between australopithecines and early *Homo erectus*. For example, one of the specimens has a cranial capacity of about 600 cm^3—bigger than the 450 cm^3 average of australopithecines but smaller than WT 15000's 900 cm^3.[14] Given the transitional appearance of these skulls and that australopithecines and early *Homo erectus* have radically different body builds, scientists were eager to find out what the Dmanisi bodies looked like. Would they have looked like Lucy's, WT 15000's, or something in between?

David Lordkipanidze of the Georgian National Museum and his colleagues analyzed the skeletons of three newly excavated Dmanisi hominins—one large and two small—as well as a partial skeleton from an adolescent (who may have been only slightly older than WT 15000).[15] They found that the Dmanisi people were transitional in body build as well as brain size. Adult heights were between four feet, ten inches, and five feet, five inches—taller than Lucy's approximate three feet, six inches, but shorter than WT 15000's projected adult height of around six feet. On the other hand, the legs of the Dmanisi individuals were long, like those of modern humans, and their limb proportions and spines looked more modern than did those of australopithecines. Their big toes were also rigid and non-opposable, like those of modern humans, so they would not have been able to cling with their feet. The feet may not have been completely modern, however. According to Lordkipanidze, the Dmanisi hominins walked with their feet turned in slightly, and with pressure more evenly distributed across all of the toes than with striding humans, although this interpretation has been questioned.[16] Nevertheless, the overall conclusion is that these early hominins were capable of long-distance travel.

Although they obviously were accomplished bipeds, the Dmanisi hominins had shoulder blades that were angled upward, compatible with climbing in trees—just like the Dikiki baby's shoulder blades 1.5 million years earlier. Significantly, this last vestige of tree climbing changed rather abruptly. Fewer than

200,000 years after the Dmanisi hominins lived, WT 15000's shoulder blades no longer angled upward where the arms attached.[17]

If Dmanisi mothers still used their apelike upper arms to climb trees, some of their little ones, lacking opposable big toes, likely became targets of natural selection as they fell off their climbing mothers. And any hominin group that allowed mothers with babies to sleep by themselves on the dangerous ground would have been selected against. So it may have been mothers with increasingly helpless babies who finally stopped sleeping in trees, propelling the rest of the group out of their tree nests and onto the ground. After the transition to ground sleeping and the loss of clinging ability that accompanied it, early hominins needed both added protection against predators and new ways to care for their young, and the solutions to both problems were social. We can safely conclude that by approximately 1.8 million years ago, the date of the Dmanisi hominins, the natural experiment in which mothers began regularly putting down their babies must have been well under way.

PUTTING THE BABY DOWN

Frank Marlowe has done extensive research on the hominin shift to ground sleeping.[18] He describes the daily rhythm of chimpanzees as "feed as you go (and sleep where you are)," and the very first tree-dwelling hominins would have followed the same rule. When changing weather patterns depleted forests, food became more widely distributed across the land. For everyone to get enough to eat, our ancestors would have dispersed into small groups that traveled during the day. However, when they began sleeping on the ground, they would have needed the safety of a large group, especially at night. As Marlowe points out, this meant that small foraging groups would have begun meeting at a central place to sleep, and this is exactly what some savanna baboons do as a defense against predators.[19]

Marlowe believes that at first our ancestors simply reconnected to sleep.[20] However, they eventually must have started bringing collected food back to their central sleeping places to share with others. If he is right, Central Place Provisioning, as he calls it, would have drastically changed hominin social life—including child care.[21] But how exactly did this crucial transition occur? Marlowe's answer is based on his extensive fieldwork among the Hadza, hunter-gatherers who live in northern Tanzania around Lake Eyasi.

Only about 1,000 Hadza are left, and Marlowe visits their camps each summer. Hadza men collect a fruitlike gourd with an edible pulp (called baobab) and honey, and use bows and arrows to hunt mammals and birds, often alone. The women, on the other hand, go on treks for several hours each day to dig tubers, gather berries, and collect baobab as well. Often they forage with a few other women and some older children, and they carry their nursing infants tied to their backs. Similar to the average for other human foragers, Hadza women have babies about every three years.[22] This means that they are likely to have a new nursing infant when their previous child is still a toddler, and traveling with both children is very difficult. According to Marlowe, "Human mothers with carrying devices can tie infants to their backs while foraging but weaned children are too big to carry and too young and small to keep up. Women can solve this problem by leaving weanlings behind in a central place if others will stay behind to look after them."[23]

Consequently, weaned children from two to more than four years of age spend more time in Hadza camps than any other age group. Other camp denizens include elderly or disabled individuals, those who are temporarily injured or ill, and older children. Marlowe's graduate student, Alyssa Crittenden, is studying babysitting practices among the Hadza.[24] She has found that babysitting (or *allomothering*) is fairly cooperative, although relatives provide the most care, much of which entails holding and carrying infants. Grandmothers are important babysitters, and their help may contribute to their children's as well as their grandchildren's survival.[25] Fathers and unrelated helpers also provide child care. Even young children, including boys, engage in babysitting. Ultimately, Hadza mothers must leave their weaned children in camp, where there are willing babysitters, and this appears to be an important reason why central places exist.

FIGURE 4.3. Hadza boys tend to be conscientious babysitters. *Photo courtesy of Alyssa Crittenden.*

FIGURES 4.4. Alyssa Crittenden does research on babysitting among the Hadza in Tanzania. *Photo courtesy of Alyssa Crittenden.*

Marlowe believes that hominin mothers would not have begun to leave their offspring in central camps until our ancestors became successful enough at hunting and gathering to bring extra food to the babysitters. Food surpluses would also have provided added calories needed for more pregnancies, as well as the means for weaning babies earlier (a paste made from baobab is an important weaning food for the Hadza).[26] Campwide food sharing is common among groups of people who still make their living by hunting and gathering, and Marlowe thinks an ability to transport surplus food back to a central place must have been at the heart of the evolution of food sharing.

The Hadza use cloaks or slings made of animal skins (which Marlowe calls karosses) to transport babies, and they probably became widely available only after hominins were successful at hunting and gathering. Baby slings clearly were invented to compensate for the evolutionary loss of infants' ability to cling unaided to their traveling mothers, but Sarah Blaffer Hrdy notes that their invention also would have marked the invention of the "carryall," allowing gathering mothers to transport not just their babies, but also food in their slings.[27] Wherever I travel in Africa, I see women carrying babies—and other things—in slings. I am amazed at how strong these women are and how content their babies seem in the slings. As discussed, babies in traditional cultures cry less frequently, probably because of the sustained physical contact in the slings of caregivers. Mothers in industrialized communities also use baby slings, of course, but to a much lesser degree. Instead of spending time in baby slings and experiencing the rhythm of their mothers' steps, our infants are taken for car rides or placed in rockers to encourage tranquility and sleep.

As Marlowe has documented, Hadza mothers "take their nursing infants with them foraging, carrying them in a kaross on their backs. A woman will keep her infant on her back even while digging tubers, occasionally swinging it to the front to nurse."[28] I asked Marlowe whether women ever remove their infants from their backs and put them down when they are digging for tubers. To my delight, he responded that he has seen Hadza women do just that.

In light of what is known about the primate need for constant contact between mothers and infants, it seems clear that early hominin mothers, like contemporary Hadza women, would have traveled with their nursing infants. However, even before our ancestors ceased sleeping in treetops, they may have lost their grasping big toes (possibly by 3.3 million years ago), which would have begun to inhibit infants' ability to cling to their mothers. Subsequent pressures from increased brain size and narrowed birth canals would have made traveling with increasingly helpless babies an onerous task. By about 1.6 million years ago, hominins had probably invented a solution still used today by Hadza and other mothers around the world: baby slings.

Although they probably coaxed older siblings or others into helping, before baby slings were invented, ancestral mothers of helpless infants almost certainly would have put their babies down to dig for tubers, harvest berries, or just gather a handful of flowers. After that, it would not be long before they discovered that soothing, cooing voices reassured fussing babies. Such melodic vocalizations, I believe, led to the first baby talk. In the next two chapters we'll explore the crucial role of modern baby talk in helping infants worldwide learn to speak and—by extension—its prehistoric role in kindling the first sparks of language.

CHAPTER 5

THE
SEEDS OF
LANGUAGE

INTERY, MINTERY

Intery, mintery, cutery corn,
Apple seed and apple thorn;
Wire, brier, limber-lock,
Five geese in a flock,
Sit and sing by a spring,
O-u-t, and in again.
—Mother Goose

A S SOON AS AN infant is placed in my arms, I am compelled to bounce gently, pat its back, and coo. It seems many women share the same impulse. Baby talk is universally used among women, many men, and even very young children. In turn, children develop various language skills through their exposure to motherese. This chapter examines motherese from a cross-cultural perspective and explores how it helps babies learn language—even before they are born.

The idea that baby talk helps babies learn language is controversial, partially because motherese can be defined in various ways. In its narrowest sense, motherese refers to the special singsong way adults talk to infants, and it is variously called "musical speech," "baby talk," or, more dryly, "infant-directed speech."[1] Experiments have shown that infants prefer motherese over the less melodious speech directed toward adults and that this preference increases during the first several months of life.[2] Although adults use motherese until their infants are about three years old, they use it most intensively with three- to five-month-old infants.[3] Motherese is exaggerated, stressing certain syllables within words, and certain words within sentences:

Aren't YOU a nice BAby? Good GIRL, drinking all your MILK.

Look, look, that's a giRAFFE. Isn't that a NICE giRAFFE?

DOGgie, there's the DOGgie. Ooh, did you see the lovely DOGgie? [4]

Compared with speech directed toward adults, motherese is slower and more repetitious, has a higher overall pitch, uses a simpler vocabulary, and includes special words such as "doggie" and "bye-bye." It is usually composed of short, straightforward sentences and contains concrete words that describe the child's immediate environment. Baby talk contains a high proportion of questions, and it increases in complexity as infants grow, because mothers automatically tailor their speech to their babies' comprehension level.

Above all, motherese is known for its musical quality, or prosody, which provides the melody or tone of voice in adult speech, coloring it with nuance and revealing emotions. It can also be a powerful motivator and memory aid—in the children's alphabet song, for example.[5] Prosody should be familiar to anyone who has used baby talk at some point. If not, motherese can be observed at a grocery store, cafeteria, playground, or other locations where mothers are likely to be with their infants.[6]

Motherese is not only verbal—it also encompasses facial expressions, body language, touching, patting, caressing, and even laughter and tickling. This broader definition takes into account the complexity of social communication. As we speak, we gesticulate, move our shoulders, and provide added meaning with facial movements. Much of body language is unconscious and intuitively understood by listeners. Though some linguists view gestures as secondary to speech, in fact body language is a powerful source of communication that in many ways speaks louder than words. Our ability to rapidly process fleeting facial expressions, for example, explains why we often can detect lies. Because mother-infant communication involves body language as well as vocalizations, some define motherese as multimodal and focus their research on facial expressions and gestures.

There is also debate about exactly who engages in motherese, and I do not mean to imply that mothers alone use it. In the United States, for instance, I have observed not just mothers using baby talk, but also fathers, siblings, children (some very young), aunts, uncles, grandparents, and strangers. Some even use motherese to address pets and foreigners. This book focuses heavily, but not exclusively, on mothers because its evolutionary perspective emphasizes comparisons with wild chimpanzees. Chimps, of course, are intimately connected to their mothers and do not recognize their fathers.

Another important part of defining motherese is acknowledging that infants take part in the process. Mothers' communications with their infants, vocal and otherwise, are molded by the infants' responses. Babies do not passively await parental input; they take an active role in programming their own nervous systems, ultimately leading to the development of language.

Motherese does more than just prompt language skills. Just like the lullabies and playsongs sung to infants all over the world, motherese is melodic and initially conveys meaning that is emotional rather than linguistic.[7] From birth, babies everywhere are predisposed to respond to melodic vocalizations that alert, soothe, please, and occasionally alarm them.[8] Because their nervous systems are so receptive to it, baby talk contributes to infants' emotional regulation and eventually to their social maturation. Although some linguists claim the functions of motherese stop there, mounting evidence shows that such exposure also helps infants learn language in a sequential, age-appropriate way.[9] So *how* exactly does motherese accomplish these impressive feats?

TILLING THE SOIL

Linguists often look at motherese from a top-down perspective. How, they ask, can mothers' melodic cooing possibly help infants grasp grammar (the rules that define a language), syntax (rules for arranging words into phrases and phrases into sentences), recursion (the inclusion of phrases within phrases, ad infinitum), and semantics (the meaning of words and phrases)? From this point of view, their bewilderment seems reasonable. The problem, however, is that they are viewing language development and baby talk as two separate entities—as if they were apples and oranges—when the acquisition of language and motherese actually are related to each other like apples and apple *seeds*.[10]

One way to learn about how language emerges in children (developmental psycholinguistics) is to examine their prenatal sound experiences. Scientists do this by placing tiny microphones outside the uteruses of (extremely dedicated) pregnant colleagues, or by using ultrasound to measure changes in fetal kicks and heart rate in response to specific sounds transmitted through the amniotic fluid. In experiments using these techniques, fetuses have differentiated speech from other sounds such as white noise and reacted to changes in musical styles—for instance, a switch from Muzak to Mozart. Using computers to measure how hard babies suck on pacifiers in response to various sounds (including their mothers' voices) has provided other information about newborns' auditory experiences. Another method is frequently used with older infants who have been trained to start recordings of sounds by turning their heads toward displays such as a blinking light. When they look away, the recordings stop. Therefore, these babies control how long they hear particular sounds, and preference for one type over another indicates that they can distinguish between certain sounds (for example, "ba" versus "da").[11]

Because of experiments like these, we now know that newborns recognize their mothers' voices at birth, even though postnatal speech is no longer muffled by amniotic fluid. Babies will also suck harder on pacifiers while listening to their native language than to foreign languages with different rhythms. These abilities have strong survival value because they encourage infants to pay attention to their mothers' voices and play a vital role in early mother-infant bonding.[12] Studies using the above techniques also provide a fascinating picture of what late-term fetuses do as they wait to be born:

> From as early as twenty weeks gestation, the hearing system of the fetus is sufficiently developed to enable it to begin processing some of the sounds that filter through the amniotic liquid. The fetus's world is filled with a cacophony of gurgles and grumbles from the mother's body, along with the constant rhythm of her heartbeats. These noises provide early auditory stimulation. But most stimulating of all are the filtered sounds of language. From the sixth month of gestation onward, the fetus spends most of its waking time processing these very special linguistic sounds, growing familiar with the unique qualities of its mother's voice and of the language or languages that she speaks. It also becomes sensitive to the prosody—the intonation of sentences and rhythm patterns within words—that structures her speech. In its last three months in the womb, the fetus is busy eavesdropping on its mother's conversations. . . . The newborn comes into the world prepared to pay special attention to human speech, and specifically to his mother's voice. These earliest intrauterine experiences prime the newborn for linguistic input.[13]

Just as adults learning a foreign language begin to grasp the meaning of words and phrases long before they can produce them, the complicated process of learning a language begins in fetuses with simple listening. Of course, fetuses and newborns don't understand the meaning of speech. Instead, they absorb the intonations, rhythms, stresses, and melodies of their mothers' voices.

Ultimately, a baby's first step on the journey to learning language is to become sensitive to its melodic features. It is no wonder, then, that prosody-laden motherese becomes an important vehicle for understanding and eventually producing speech.

If you've shared my experience of struggling as an adult to learn a foreign language, you will remember that, at first, speech seems too rapid to grasp and is, for all practical purposes, meaningless. A beginning listener is unable to tell where one word ends and the other begins. With persistence, however, the student begins to get the feel (rhythm) of the language and learns individual words that then stand out in an otherwise meaningless speech stream. With more time, the listener begins to grasp small chunks of the linguistic flow, and so on. The language learner, then, must learn to parse the speech stream into recognizable units of words and phrases while attempting to decode their meaning. This is not an easy task even for grown-ups who are familiar with the concept of language and can often use vision and context to interpret meaning. For example, a non-Spanish speaker understands when he or she can *see* a speaker raise an empty beer mug in a gesture to the bartender while saying, "Otra cerveza, por favor."

The fetus, on the other hand, has no visual aids. She relies only on muffled sounds. By the time the fetus reaches the third trimester, her job is to learn to recognize the rhythms, stresses, intonations, and melodic sweeps of the language that she will hear as a newborn.[14] Presumably, this is not true of a late-term chimpanzee fetus—evolution has sculpted this intricate pathway to language in humans alone. It is amazing to consider that our babies are born ready to learn language in concert with their mothers, while mothers have evolved to instinctively encourage their babies using the very medium (prosody) that babies worked so hard to become sensitive to in the womb. When mothers gush baby talk to their newborns, they usually have no idea that, in addition to expressing love, they are exaggerating certain parts of the speech stream and underscoring sound combinations and grammar. Indeed, because of motherese, newborns discover more easily how to divide speech into words and clauses long before they learn their meanings.

FIGURE 5.1. An infant is wired for sound, wearing an EEG net. *Photo courtesy of Laurel Trainor, director of the Auditory Development Laboratory at McMaster University in Hamilton, Ontario, Canada.*

Scientists study infant speech perception in various ways, including fitting infants' heads with nets studded with electroencephalogram (EEG) electrodes to measure their brain activity. For example, babies may listen to a recorded voice that repeats a single sound (phoneme) before switching to a different one. If the baby hears the distinction, a blip appears in the recorded EEG. These nets may also be used to study music perception in infants. As we discussed earlier, another method for determining whether babies hear contrasting sounds involves conditioning them to turn their heads toward the source of a changed sound, which indicates that they have detected it.

Using such techniques, Patricia Kuhl and her team at the University of Washington, Seattle, discovered that six-month-olds can distinguish all the sounds in all of the world's languages—no small feat given that there are approximately six hundred consonants and two hundred vowels.[15] By the end of their first year, however, babies are on their way to perceiving especially well the sounds that are important for their native tongues (usually around forty for a given language). At the same time, their ability to distinguish foreign speech sounds decreases.[16] Japanese infants, for example, initially perceive separate sounds for "r" and "l" (as in the words "rake" and "lake") but lose the ability to hear this "foreign" distinction as they mature and become more adept at recognizing Japanese speech sounds. Like babies everywhere, they make a transition from being "citizens of the world" to becoming "culture-bound" listeners.[17] This phenomenon is consistent with Kuhl's predictions about the relationship between early infant exposure to language and an infant's future language learning.[18]

The infant's task of learning its native language is a daunting one because, unlike written language, spoken language has no obvious markers that indicate the boundaries between words. Just try reading the following sentence: Byanaverageageofninemonths infantsdiscriminatesequencesofsoundsthatoccurfrequentlyfromones thatdonot. ("By an average age of nine months, infants discriminate sequences of sounds that occur frequently from ones that do not.") English-speakers can read the sentence without breaks because they are familiar with the words and know where the spaces should be. Understanding it would be more difficult if, instead of reading, one were listening to the sentence being spoken as one long string of syllables that were equally stressed. So we see that stressing different syllables is invaluable to language comprehension. To begin to decipher separate words, an infant must become familiar with the intonations of its mother's voice to help differentiate between words. Of course, it would be even harder to decipher the above sentence if it contained only unfamiliar words, which gives us a better idea of the challenge facing preverbal infants.

Long before they can speak, babies become sensitive to the frequencies at which combinations of syllables occur and how they differ within and across word boundaries. Kuhl uses the example of the phrase "pretty baby" and notes that among English words, the likelihood of "ty" following "pre" is higher than the likelihood that "bay" will follow "ty." Thus, with enough repetition, babies begin to understand that "pretty" is potentially a word, even before they know what it means.[19] Prosodic cues embedded in language also help. In conversational English, for example, a majority of words are stressed on their first syllable, as in the words "monkey" and "jungle."[20] This predominantly strong-weak pattern is reversed in some languages. By the time an English-learning infant is seven and a half months old, he spontaneously perceives words that reflect the strong-weak pattern, but not the weak-strong pattern. So when infants hear "guitar is" they perceive "taris" as a unit because it begins with a stressed syllable.[21] As we've learned, the singsong quality of motherese is enormously helpful for marking the divisions between words.

Significantly, infants in Kuhl's laboratory who were best at perceiving speech sounds at seven months of age also scored higher when they were older on language tests that measured the number of words they could say and the complexity of their speech.[22] Although the developmental pruning of the speech sounds an infant can perceive paves the way for recognition and comprehension of the words in her native language, speech discrimination is not the only critical factor in early language learning. The clarity of motherese that normal babies are exposed to is associated with their development of speech discrimination skills.[23] This fact has been associated with babies' social interest in the kind of speech caregivers direct toward them.[24] It is not surprising, then, that motherese does not appear to be as beneficial for linguistic development in "children with autism, who lack a social interest in communication."[25]

A linguist once protested to me that motherese had nothing to do with acquiring language and was only an emotionally loaded mechanism for strengthening the mother-infant bond. Without a doubt, this is one of the important functions of motherese. But this does not negate that motherese also helps babies learn language, as shown by an important study comparing various qualities of mothers' voices as they spoke to their six-month-old infants, their cats or dogs, and other adults.[26] The experiment found that mothers do speak to their babies and pets—but not to other adults—with heightened emotion expressed in pitch, tone of voice, and speech rhythm.

But the study also discovered a difference in how women speak to their babies and pets. Mothers unconsciously exaggerate vowels when addressing their infants but not with their pets, which shows that, in addition to expressing love, baby talk provides fundamental building blocks (in this case, knowledge about vowels) that babies need to learn their native tongues. And this finding applies not only to American mothers. English, Russian, Japanese, and Swedish mothers also exaggerate vowels when addressing their infants, but not when addressing other adults.[27] How much mothers exaggerate their vowels also is strongly associated with infants' speech perception: the more exaggerated a mother's vowels, the better her infant's performance.[28] (One study suggests that computers also find it easier to learn vowels presented in motherese rather than adult-directed speech.[29])

The degree to which vowels are emphasized is only one difference between the way mothers speak to babies and the way they speak to their pets. A relatively large proportion of mothers' words to their infants (and not to their dogs or other adults) is instructive. That is, they effectively "point things out."[30] For example, a mother might say to her infant, "That's a kitty-cat."[31] Similarly, mothers ask their infants, but rarely their pets, tutorial questions, such as "What color is this?" In addition, mothers do not usually take both parts in "conversations" with their dogs, as they sometimes do with young infants. For example, a mother might ask, "What's the matter with your knee?" and then answer, "Oh, Baby has a boo-boo."

Of course, after an infant is born, mother-infant communication involves the newborn's feedback, as each partner responds to the other's behaviors. If you stick your tongue out at a newborn, for example, he is likely to imitate the gesture.[32] Andrew Meltzoff of the University of Washington has shown that twelve- to twenty-one-day-old infants can imitate at least four adult gestures: tongue protrusion, lip protrusion, mouth opening, and finger movement.[33] Significantly, this suggests that there is an innate link between perceiving acts and producing them.[34] As they mature, babies correct their imitative behavior and mentally store models that they later imitate from memory. Surprisingly, Meltzoff's research also reveals that infants not only are born with the ability to imitate others but also know when others are imitating them. Reciprocal imitation, in fact, may be the first building block for developing the ability to communicate:

> Human beings do not only imitate. They also recognize when they are being imitated by others. Such reciprocal imitation is an essential part of communicative exchanges. A listener often shows interpersonal connectedness with a speaker by adopting the postural configuration of the speaker. If the speaker furrows his or her brow, the listener does the same; if the speaker rubs his chin, the listener follows. Parents use this same technique, however unconsciously, in establishing intersubjectivity with their preverbal infants.[35]

Infants pay close attention to faces, and by the time they are four months old prefer to listen to speech sounds that are coordinated with visual images of appropriately shaped mouths.[36] In other words, they have learned to attend to mouth shapes that correspond to what someone is saying. Mothers' facial expressions reinforce motherese, while their infants' faces and vocalizations indicate whether they are paying attention and how they are feeling. These and other co-regulated behaviors create rapid exchanges that are a kind of mother/infant conversation.[37] Mothers begin these exchanges by unconsciously establishing eye contact with their infants and using baby talk. As they realize their babies are responding by jerking their limbs, cooing, and gurgling, mothers begin taking turns with them. Although such conversations are initially one-sided, they prepare infants for the turn-taking that will be so vital to future interactions.[38]

So with the help of motherese and their exquisite observational skills, infants become sensitive to the patterns of speech sounds in their native languages. As they mature, they use these cues to learn other aspects of language, including the nature of syllables, words, phrases, and sentences.[39] (We'll learn more about this process in Chapter 6.) It would be a mistake, however, to think that young babies understand the linguistic meaning of what they hear or that they acquire language precisely in a step-by-step manner. In reality, babies are little pattern-processing geniuses who develop an acute sensitivity to various aspects of their native languages, and this occurs in a complex and integrated manner.[40] However, speech perception itself is not language. Before babies can understand and use language, they must develop not only the ability to recognize speech sounds, but also the motor skills required to produce them.

FROM CRYING TO BABBLING

As we saw in the last chapter, the ability to vocally hush crying infants helped decrease the extent to which hungry predators could locate helpless little hominins—or their mothers. Despite the efforts of countless prehistoric mothers to subdue discontented infants, however, infants' crying continued to exist and even underwent its own evolution. Recall, for example, that human babies cry differently than other primate infants; that their cries can signal sickness or good health to parents, who may not choose to "invest" in ill offspring; that infant cries and maternal sensitivity to those cries evolved at least in part to suit the infant need for protection and nurturing; and that quite often crying babies simply need physical contact with their mothers. What we have not yet discussed, however, is crying's important—though not yet widely recognized—role in language development.

Crying, in fact, may be a missing link in theories about how language is learned and how it emerged in early hominins.[41] In 1985, Kathleen Wermke and Werner Mende became interested in developing models for diagnosing certain brain dysfunctions and diseases by recording and analyzing infants' cries. At that time, most researchers thought crying bore no relationship to the emergence of language in young children. Wermke and Mende soon discovered, however, that crying undergoes dramatic changes for some months after birth. In fact, the complex development of infant crying was getting in the way of discovering patterns that could indicate specific illnesses. As Wermke and Mende remember it: "This completely changed our view on the infant cry as a simple alarming bio-siren. A siren would lose its alerting character if it would continuously change its signal signature. Another explanation was found—interpreting cry melody development as systematic development toward language by applying a modular composition principle using melody arcs as building blocks."[42]

A melody arc is a single rise, then fall of pitch in a sound an infant produces in one expiration of air. Wermke and Mende discovered that the single-arc melodies infants produce in their first weeks of life develop over time into increasingly complex cries that contain numerous melody arcs. Long before infants begin babbling or using words, repetitions of single-arcs lead to complex cries of two, three, or more melody arcs within single breaths.[43] Interestingly, Wermke and Mende found that the double-arc melodies of German-born infants were accented in a way that could provide the basis for producing the accents typical of German speech.[44] They also noted that, in addition to producing more complex melodies, infants soon could produce highly stable sounds, comparable to the singing of intended musical notes. Wermke and Mende viewed the development of complex melodies in young infants' cries as a first step in vocalizing emotions and needs.[45]

Speech sounds do not come from the vocal cords alone. Instead, the waves of air the cords produce are molded into the components of speech (vowels, consonants, words) by parts of the anatomy above the voice box—the throat, tongue, mouth, and teeth. I like to think of talking as the process of expelling pieces of chopped-up air from the mouth. In other words, the entire vocal tract molds vibrating streams of air the vocal cords produce into specific speech sounds. This process occurs extremely rapidly. To experience this, put your hand on your throat, speak aloud, and notice what your tongue and mouth do.

From a neurophysiological perspective, the production of sound waves from the vocal cords (phonation) and their articulation into speech sounds by the vocal tract are independently controlled, so these processes must become tuned to each other during development. Wermke and Mende's ongoing research on crying has shown that the two systems begin to synchronize by the time infants are about three months old, and that by the time they are four or five months old, babies make many vowel-like sounds and near-syllables. When infants babble at around seven months, they can produce well-formed syllables.[46] So the crying in pre-babbling infants undergoes a "stormy development" that trains their vocal cords and vocal tracts to perform the incredibly complicated and rapid gymnastics required for intentional speech.[47]

Christopher, who produced the cries shown in Figure 5.2, underwent such a development. He produced melodic cries at two months, started babbling vowels by four months, and began producing consonant-vowel syllables between nine and ten months. By eleven months, he was uttering complex melodies and producing his first words, *Mama* and *ein Teddy* (a teddy bear). By then, Christopher was fascinated by books and "read" them on his own by turning the pages and imitating the prosodic patterns

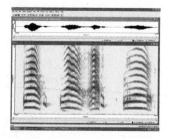 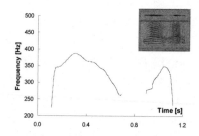

FIGURE 5.2. Cries of Wermke and Mende's subject Christopher when he was two months old. The spectrograms on the left show three subsequent cries that alternate between single-arc, double-arc, and single-arc melodies. The melody of the central double-arc cry is shown in the contours on the right, which are separated by a short break. Such breaks are prerequisites for later babbling and are already acquired at this early stage. *Image courtesy of Kathleen Wermke.*

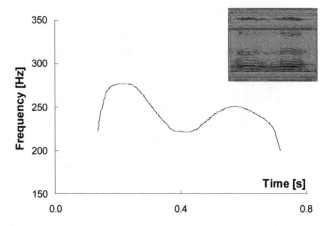

FIGURE 5.3. The contour for one of Christopher's two-syllable words using the same melodic building blocks as those used during early babbling and crying. *Image courtesy of Kathleen Wermke.*

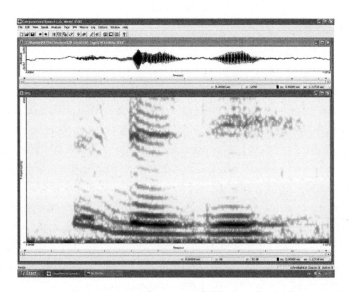

FIGURE 5.4. Spectrogram for one of Christopher's first words, *ein Teddy* (a teddy bear). *Image courtesy of Kathleen Wermke.*

FIGURE 5.5. Christopher "reads" a book by imitating his mother's prosody. *Photo courtesy of Kathleen Wermke.*

his mother had used when she read the book to him several days earlier.

Wermke and her colleagues have taken their research a step further in showing how infants' changing cries feed language development.[48] They first evaluated language skills in thirty-four healthy German children when they were two and a half years old. They recorded the number of words each child produced and assessed each one's comprehension and production of words and sentences. Then the children were separated into two groups based on their language skills. Twenty-four were in the normal group, while ten showed delayed language development. (This number was unusually high because toddlers who were likely to experience delays in language development had been specifically recruited for the study.) Wermke and her colleagues then analyzed and compared cries that had been recorded for each infant in hospitals and at home at two-week intervals from birth until they were four months old. The results were startling. Infants with significantly fewer complex melodies in their cries during their second months were five times more likely to be delayed in their language development by the time they reached two and a half years of age. These findings indicated that the second month of life is a critical period for attaining the melodic skills necessary for normal language development, so crying plays a key role in developing language.

Because of my interest in motherese, I asked Wermke and Mende whether maternal behavior influences the melodies of infant cries. Although they have not yet conducted a scientific study related to this particular question, they believe mothers do indeed impact their infants' crying melodies. Wermke and Mende note that infants are recorded as they begin to fuss and cry in bed. The babies usually try to make eye contact with their mothers, and then start crying more intensely. The mothers then speak to their babies and pick them up, at which point the babies change their crying melodies to different patterns.

As noted, Wermke and Mende's crying research is on German infants. Do the crying melodies of infants exposed to different languages vary? Should we speak of German crying, Russian crying, and English crying as unique? Do infants in cultures with click languages cry in different melodies that help them learn their language? These are fascinating questions that, with luck, will one day be answered.

Meanwhile, the studies of Wermke and Mende have shown that an ability to perceive, produce, and process melodies is vital for language development in infants. Wermke and Mende also see melody as a kind of semantic filter that extracts life-relevant information out of the complex parental speech stream.[49] In fact, they believe melody is at the root of oral languages and would therefore have been extremely important for the emergence of language during prehistory. This idea is not at all far-fetched. Recall that the melodic components of motherese are especially appealing to babies and that the singing of lullabies, which shares these components, occurs across all cultures. Wermke and Mende's twenty-plus years of research have also clearly demonstrated that crying melodies are precursors to babbling—and babbling, as we'll see, is essential to learning speech.

In their first few months, babies warm up their vocal cords by producing nonspeech sounds such as cooing, fussing, and blowing bubbles and raspberries. By four months, they also engage in vocal play and begin articulating vowel-like sounds and near-syllables. By that time they have also developed complex melodic cries that pave the way for babbling.[50]

Although I am captivated by babies of all ages, I think they might be most adorable around seven months old—when they smile, babble a sequence of nonsense syllables, pause, and look at you as if to say, "Well, isn't that so?" By this point, infants produce well-formed consonant-vowel combinations such as "ba," "da," and "goo." The repetition of these combinations allows babies to babble in the rhythms and melodies of their native languages ("goo-goo baba dadada"). At least two things occur during babbling. First, infants continue learning how to move their throats, mouths, tongues, and lips to produce the speech sounds they have been hearing. The reason babbling seems so speech-like is that its prosodic features reflect those of the native language. For example, French babies use the predominant French pattern of final syllable lengthening when babbling, but English-learning babies do not.[51] Babbling also represents an important link between an infant's ability to produce the sounds of its native language and to speak its first real words, which we will discuss later.

The suggestion that babbling is an important milestone in learning language is consistent with the discovery that Finnish, English, and French babies use the left sides of their brains when babbling but not when making other sounds, just as adults do when speaking.[52] In most societies, of course, babbling unfolds in social contexts that involve mother–infant imitation, turn-taking, and baby talk. Nevertheless, some anthropological linguists have claimed that there are a few cultures in which infants learn their languages without being exposed to motherese. Given what we've discussed so far, this is a fascinating and surprising suggestion. But is it true?

Is Motherese Universal?

Social scientists have long sought to discover the common features that characterize baby talk across cultures. About thirty years ago, Charles Ferguson identified approximately thirty widespread characteristics of baby talk across twenty-seven Indo-European, African, and oceanic languages.[53] Predictably, he found that the most notable characteristics of speech addressed to young children included overall higher pitch, exaggerated intonation, and slowed speed—prosodic features. Ferguson concluded that a baby-talk "tone of voice" is universal to human language behavior. He gave three possible reasons: First, speakers might be imitating the pitch infants produce, which is much higher than that of adults because of the size and shape of their vocal tracts. Second, baby talk corresponds to infants' perceptual sensitivities and therefore attracts their attention. Finally, the prosody of baby talk helps infants segment the speech stream, providing hints about the structure of their native languages. As we have seen, all of Ferguson's ideas have been confirmed in recent years.

Tone of voice, however, is not the only widespread feature of motherese. As we will see in Chapter 6, baby talk everywhere tends to have words composed of duplicated syllables (mama, papa, boo-boo); a special vocabulary that includes a high proportion of names for body parts and functions, food, animals, and games (peekaboo, patty-cake); and special constructions, such as the compound verb "go bye-bye." When addressing infants, caregivers across cultures usually speak in present tense, ask many questions, and use tags such as "okay?" and "hm?" Sentences addressed to infants are shorter and simpler than those spoken to adults, and baby talk often includes repetition. As babies grow and become more competent in language, caregivers around the world modify their speech accordingly. Significantly, Ferguson recognized that baby talk is universal but that the composition and frequency of its features vary somewhat from culture to culture.[54]

This is an important observation, because the universality of motherese has been challenged based on assertions about a few exceptional cultures. For example, in writing about the "baby talk register," one anthropologist observed: "We now know that the process of language acquisition does not depend on this socio-linguistic environment. Western Samoan, Kaluli New Guinea, and black working class American children are not surrounded by simplified speech . . . and yet they become perfectly competent speakers in the course of normal development."[55]

Although there are others, the three cultures mentioned in the above passage are most often cited as evidence that motherese is not universal. For this reason, I decided to investigate these cultures to see if their babies really do grow up without being exposed to any motherese. What I found was both compelling and instructive about how much language is shaped by nature and nurture alike.

The Kaluli are a small, nonliterate, egalitarian society living in a tropical rain forest in the southern highlands of Papua, New Guinea. After she lived among the Kaluli in the 1970s, anthropologist Bambi Schieffelin reported that the Kaluli did not use baby talk because they believed "that to do so would result in a child sounding babyish."[56] Schieffelin documented interesting differences between social interactions involving infants in white middle-class America and the Kaluli. These differences arise partly because the Kaluli live in extended families in large semi-partitioned dwellings, where conversations tend to involve multiple parties rather than just two people. As a result, Kaluli mothers did not speak directly to their infants as much as some American mothers do. Although the Kaluli may not have special baby talk words, Schieffelin's work suggests that motherese is indeed present among the Kaluli and that their child-care practices are consistent with those we discussed earlier. For example, "In the first two years, children spend almost all of their time with their mothers and siblings. Mothers, who are the primary caregivers, are attentive to their infants and physically responsive to them. When infants cry while being carried in the net bag, mothers gently and rhythmically bounce up and down to soothe them. Their movements are often accompanied by repeated sounds of "Shh," and the whole activity is called *hɛnulab* "persuade, buy off, distract."[57]

Schieffelin noted that infants were always in physical contact with their mothers, either sleeping in net bags next to their bodies or being held in their arms. Mothers never left their young infants alone and rarely left them with others. In the first few months of babies' lives, their mothers greeted them by name and used "expressive vocalizations."[58] Mothers did not, however, gaze directly into their infants' eyes, which is normal for interactions between adults. Although some claim Kaluli mothers and infants do not engage in proto-conversational interactions, there are culturally acceptable ways to expose Kaluli babies to such dialogues through baby-talk (high-pitched) tones:

> Within a week or so after a child is born, Kaluli mothers act in ways that seem intended to involve infants (*tualun*) in dialogues and interactions with others. Rather than facing their babies and engaging in dialogues with them in ways many English-speaking mothers would, Kaluli mothers tend to face their babies outward so that they can be seen by and see others who are part of the social group. Older children greet and address infants, and in response to this mothers hold their infants face outward and, while moving them, speak in a special high-pitched, nasalized register (similar to one that Kaluli use when speaking to dogs). These infants look as if they are talking to someone while their mothers speak for them.[59]

As infants grow older (six to twelve months old), adults address them with greetings, rhetorical questions, directions, and simple "one-liners."[60] When babies babble, "adults and older children occasionally repeat vocalizations back to the young child, reshaping them into the names of persons in the household or into kin terms."[61] Mothers respond to their toddlers' shrieks upon being harassed by older children by producing (as an interpretation of their babies' cries) utterances that translate as "I'm unwilling."[62] The Kaluli practice these aspects of motherese, despite believing that language does not begin until children use the words for "mother" and "breast."[63] From then on, however, infants learn about language by continual requests to repeat their mothers' utterances with a special word (*ɛlɛma*) that means "say like that."[64] Kaluli mothers also correct their children's pronunciation and word choice. All of these observations show that the Kaluli do have motherese, although it is modified for cultural norms, just as Ferguson suggested three decades ago.

What about the Western Samoans, who also supposedly lack motherese? Anthropologist Elinor Ochs studied language acquisition on the island of Upolu in Western Samoa in the late 1970s. She reported that traditional Samoan homes have no walls and, like those of the Kaluli, conversations tend to involve multiple parties rather than just two individuals. Accordingly, infants are usually held outward to face the social group. The Samoans are a highly socially stratified society, and mothers normally direct lower-ranking people to care for and speak to their still lower-ranking infants. According to Ochs, "Caregivers do not speak in a dramatically more simplified manner to very young children . . . because such accommodations are dispreferred when the addressee is of lower rank than the speaker."[65]

Although Ochs's remarks have been interpreted widely to mean that Western Samoans lack motherese, many of her observations, in fact, show just the opposite.

Babies from birth until about six months are referred to as *pepemeamea* ("baby thing thing") and exposed to the prosody that characterizes motherese. For example, "infants are sung to and cooed over to distract them from their hunger or to put them asleep or simply to amuse them."[66] During the first half-year of life,

> the infant spends the periods of rest and sleep near, but somewhat separated from, others, on a large pillow enclosed by a mosquito net suspended from a beam or rope. Waking moments are spent in the arms of the mother, occasionally the father, but most often on the hips or laps of other children, who deliver the infant to his or her mother for feeding and in general are responsible for satisfying and comforting the child. . . . Language addressed *to* the young infant tends to be in the form of songs or rhythmic vocalizations in a soft, high pitch . . . once a child is able to locomote . . . the tone of voice shifts dramati-

cally from that used with younger infants. The pitch drops to the level used in casual interactions with adult addressees. [67]

As they continue to mature, children learn language with methods similar to those reported in the Kaluli, including numerous requests to repeat utterances.[68]

Working-class African-Americans have also been cited as lacking motherese.[69] While doing research in the 1970s in an African-American community of low socioeconomic background in South Carolina, linguistic anthropologist Shirley Brice Heath observed that adults in the community did not speak slowly to their infants or use a special pitch. They also did not simplify their infant-directed speech by substituting easier sounds for more difficult ones or using special baby-talk words. However, new research by Monique Tenette Mills suggests otherwise. Mills studied the speech of four African-American women from Columbus, Ohio, to their infants or grandchildren.[70] The infant-directed speech was higher-pitched and more exaggerated than their speech to other adults. When addressing babies, the women in Mills's study simplified their vocabularies and repeatedly interjected questions. Diminutives such as "night-night," "ea-eat," and "no-nose" were frequently used, and baby-talk words were substituted for others, such as "sheepy" for "sleepy" and "binky" or "paci" for "pacifier." These women were clearly using motherese.

Mills offered several reasons why her results may have been so different from Heath's: the way data were collected differed between the two studies, and results may have been influenced by cultural differences between the South Carolina community Heath studied in the 1970s and the northern city of Columbus, Ohio, that Mills studied. Mills also suggested that the fact that she is African-American and speaks African-American Vernacular English (or AAVE) may have increased her subjects' comfort level.

The above studies show that motherese is, to some degree, culturally constrained. If it is considered impolite to make direct eye contact, then that taboo likely will be incorporated into motherese. If the conversations normally take place between more than two people, then the motherese an infant hears will probably come from various sources. If mothers are socially un-comfortable talking to lower-ranking individuals, such as infants, then they may arrange for lower-ranking individuals to speak to their babies. When speculating about motherese, many anthro-pological linguists focus on the conscious teaching of language to toddlers, a very late stage in language acquisition. Prenatal and in-fant exposure to the prosody of language teaches babies the rhythms of their native language, regardless of whether mothers are consciously attempting to instruct. And as we have seen, cul-tures that supposedly lack motherese nevertheless expose young infants to rocking, cooing, singing, and baby-talk tones of voice. Such prosody, of course, is one of the most important and wide-spread manifestations of motherese.

There is one report, however, of a culture in which infant-directed speech lacks the high pitch that is almost universally as-sociated with motherese. Linguist Clifton Pye studied infant-directed speech in Quiché-speaking mothers from Guatemala's western highland region in 1977. Pye's sample included only three mothers and toddlers, and the youngest was twenty-two months old. Contrary to expectations, the mothers' pitches re-mained the same or decreased slightly when they addressed their infants, which led Pye (and others) to reject the hypothesis that motherese is associated with universal features.[71] However, Quiché is a Mayan language in which, Pye speculated, high pitch may be used when addressing people of high status. If so, "any tendency to raise pitch to infants, as is done in most other cultures, would find itself in diametrical opposition to a pragmatic function which raised pitch to speakers of high status."[72] As with the other lan-guages we've discussed, upon closer examination (and as detailed by Pye), infant-directed speech in Quiché has features that are seen in other language communities, in addition to some that ap-pear to be unique.[73] Significantly, Pye concluded that sociolin-guistic rules for expressing deference in Quiché affected the modifications made in child-directed speech and that each com-munity imposed its own limitations on infant-directed speech.

Although these studies underscore that societal norms con-strain mother–infant communication, they fail to support the idea that some cultures completely lack motherese. Just as language is universal but manifests differently in different cultures—for example, Chinese is very different from English—so is motherese. As children mature and become focused on learning their native languages, the motherese they hear changes in culturally appropriate ways. In fact, Pye's finding that infant-directed speech in Quiché lacks high pitch may highlight the importance of this particular feature for acquiring language because his research also raises the possibility that Quiché infants may be slower to learn speech than some other children:

> My assistants have told me that parents do not become concerned about delays in a child's language development until the child is 3 or 4 years old. Children who do not begin to speak until 3 years of age are not given any special treatment. There are traditional remedies for language delays, however. The Tenejapa Tzeltal believe it helps to bump a child's head gently every once in a while with a large, short gourd used to keep tortillas warm. They may also give children 3 or 4 years old roasted cicadas to eat if they have pronunciation difficulties, delays in learning to speak, or problems speaking well or correctly.[74]

These and other cross-cultural studies confirm that some form of motherese is *universally* essential to infants' social maturation, emotional regulation, and linguistic development. From the moment babies are born—and perhaps even while they are still in the womb—parents should not hesitate to expose them to baby talk.

CHAPTER 6

WHAT'S IN
A NAME?

COTTLESTON PIE

Cottleston, Cottleston, Cottleston Pie,

A fly can't bird, but a bird can fly.

Ask me a riddle and I reply:

"Cottleston, Cottleston, Cottleston Pie."

—*A. A. Milne*

STRONG MOTHER-INFANT BONDS were vitally important for prehistoric infants to survive, and as a result, visual, vocal, and physical mechanisms evolved that enabled mothers and infants "to enter the temporal world and feeling state of the other."[1] Of course, this describes prosody and other features of motherese. In a workshop, Ellen Dissanayake detailed how these special behaviors could have evolved into music, dance, and mime, and she suggested that these, in turn, may have promoted bonding and cooperation—necessities for survival—among adults. What was puzzling, however, was that most of the attendees shrugged off her brilliant presentation.

Inspired by Dissanayake's work on motherese and the temporal arts, I began exploring baby talk's role in the emergence of language. I learned that some form of motherese occurs universally in human cultures but does not exist among the great apes. So motherese *itself* must have emerged at some point during human evolution. I also found that motherese helps babies worldwide learn their native languages. It seemed reasonable, then, to suggest that motherese evolved before the first language (protolanguage) and had something to do with the latter's emergence. But why did hominins first begin using motherese? I pondered this problem for some years before formulating the "putting the baby down" hypothesis, with which you are now familiar: Natural selection for bipedalism and the gradual enlargement of hominin brains resulted in the loss of infants' ability to cling unaided to their mothers' bodies.[2] Until the invention of baby slings, moms were primarily responsible for holding and carrying their babies. Hominin mothers and infants probably began interacting with soothing sounds and cries when mothers began putting their babies down nearby to forage. As we'll see below, there is good reason to believe their prehistoric vocalizations eventually led to protolanguage.

About six years after hearing Dissanayake's talk, I was invited to give a seminar for anthropologists at a California university. I was eager to share my ideas about the behaviors that paved the way for protolanguage. Although Dissanayake's talk had met with silence, mine was greeted with open scorn. One anthropologist threw down his pencil while objecting, "You haven't proven that motherese has *anything* to do with learning language. Sure, mommies are emoting affection, but there's no evidence that it has anything to do with their infants' eventual grasp of syntax, grammar, or semantics." When I protested that I was focusing on the precursors that *led* to the emergence of language, he remained unconvinced. Another anthropologist, a linguist, informed me that my ideas couldn't be true because research on the Kaluli and Western Samoans had shown that motherese was not universal (a claim I addressed in the previous chapter).

I found the response to my talk more perplexing than upsetting. I left determined to examine the cultures that allegedly lacked motherese and, further, to discover how motherese related to the evolution of the more formal aspects of language. I also began to wonder if my talk (and Dissanayake's) had been poorly received partly because we emphasized the role of women and infants—rather than men—in human evolution.

As noted, my "putting the baby down" hypothesis had been intended to explain the behaviors that *preceded* the emergence of protolanguage in our ancestors, not to provide an account of the later appearance of language itself.[3] However, linguists demanded more: "Humans have one or two unique adaptations, which include symbolic, referential units and the ability to link these in the rule-governed (and potentially infinite) structures. Some of us want to have those adaptations explained."[4] For a nonlinguist like myself, addressing these issues presented quite a challenge, so I asked my anthropologist friend Myrdene Anderson to enumerate the features thought to be universal among languages.[5]

All languages, Anderson explained, have a finite system of speech sounds (called *phonemes*) that can be clumped into vowels and consonants. They all have nouns and verbs and specific ways to modify them, which differ depending on the language. All have minimally meaningful units (called *morphemes*) that can be attached to words to modify their meaning. In English, for example, *s* may be added to make a word plural, *ing* to turn it into a present participle, or *ed* to make it past tense. All languages have ways to indicate relationships not only to past, present, and future time, but also to space (although some languages do not have separate prepositions per se). People everywhere use question words (in English, these include "who," "what," "where," "when," "why," and "how") and other question formations ("Did the girl see the boy?"). Some form of counting is done universally, but numerical concepts and the ways of expressing them can differ greatly. For example, an Australian language (Warlpiri) has numbers only for "one," "two," and "many." All languages, said Anderson, have suprasegmental sounds, such as tone of voice (which is universal). Raising the voice at the end of a question in English is an example of a suprasegmental phoneme.

Finally, and most significant, all languages have their own specific rules (*syntax*) for combining words into larger meaningful phrases, and phrases into sentences. Syntax and the regulations for

combining morphemes into words make up a language's grammar. Because the different combinations of words that can be
formed in any language are, for all practical purposes, infinite, and
because phrases can be embedded within phrases ad infinitum,
people everywhere can articulate an endless variety of thoughts.
Cultural taboos and practices constrain how and to what degree
these universals appear in each language, just as they constrain the
manifestations of motherese in different societies.[6] To understand
how motherese contributed to these universal linguistic rules, we
must understand more about how modern babies become linguistic beings.

So far, we have traced prelinguistic development from the
eavesdropping of prenatal fetuses through the babbling of approximately seven-month-old infants. We've learned that newborns'
exposure to motherese reinforces their innate tendency to extract
salient features from their mothers' voices and eventually molds
their ability to perceive individual words and clauses in rapid
speech streams. And young infants bring amazing computational
and pattern-processing abilities to this task. Experiments suggest
that the quality of motherese individual babies hear is associated,
to some degree, with how well they perceive speech sounds.

While infants are figuring out how to dissect speech streams,
they are also producing increasingly complex melodies when they
cry. As shown by Werner Mende and Kathleen Wermke's research,
crying eventually prompts infants to produce the syllables needed
to babble in the rhythms and melodies of their particular languages. Not only do babbling babies reproduce the elementary
speech sounds and prosody of their native languages, but they also
politely (and adorably) wait their turns in "conversations." And
babbling, like true speech, engages the left side of the brain, further indicating the fundamental relationship between babbling and
language. However, a babbling infant still is not a speaking infant.

Although we've heard a strong case for the importance of
motherese in language acquisition, we have not yet extended our
discussion beyond the babbling stage. In this chapter, we will focus on how protolanguage emerged from prehistoric motherese
and then evolved incrementally into language with both words

and syntax. This is not to say that early motherese was *directly* responsible for all the subtleties of language. However, motherese did begin a chain of events crucial for early language to emerge and, by extension, for modern language to evolve.

FROM BABBLES TO WORDS

Humans know an astounding number of words. The average American high school student knows an estimated sixty thousand of them. Linguists Steven Pinker and Ray Jackendoff observed that "the fact that words can be learned at all hinges on the predisposition of children to interpret the noises made by others as meaningful signals. . . . Much of the job of learning language is figuring out what concepts these noises are symbols for."[7] Before babies can begin to master language, they must first learn its components. This also would have been true at an evolutionary level. Before language could fully emerge, our ancestors had to invent words.

The creation of words would have been no small achievement, as they are not simply names (although, as we discuss below, the first words may have been).[8] As children learn in grammar school, different kinds of words have different functions (nouns, verbs, prepositions, and so on). Modern words are imbued with important information, such as markers of tense and indications of whether they are singular or plural. Except for proper names, words are generic. Upon learning that an object is called a ball, for example, a young child readily applies the word to other balls and assumes that other people know the word.[9] Words are "shared, organized linkages of phonological, conceptual, and grammatical structures."[10]

But how do babies first learn that words have certain meanings? As we saw in the previous chapter, well-formed babbling tends to appear when infants are about seven months old. Around the beginning of the second year, babbling becomes quite complex and babies utter their first words. In fact, it becomes increasingly difficult during this stage to differentiate between babbling and the first semblances of words. Mother-daughter collaborators Annette Karmiloff-Smith and Kyra Karmiloff ask, "When do the repeated syllables 'ma-ma-ma' become a symbol for 'mother'? Is the utterance 'ahhr' still merely a babble if the baby is pointing to a car at the same time, or is it the child's idiosyncratic yet consistent sound for 'car' that now has real referential status?"[11] The turning point (or activation of the so-called naming insight) occurs during babies' second year, when they realize the noises people make are symbolic. As adults indicate the meaning of certain syllables (e.g., "dada," "bow-wow," "go," etc.), the child discovers that a certain sound is consistently associated with a particular person, object, or action. It is this mapping of specific meanings onto specific vocalizations that is the hallmark of words, and once babies "get it," they can begin building vocabulary. From then on, they must figure out what words mean and how to use them to get what they want.[12] To determine the meaning of words, they pay attention to the intonation patterns, stresses, and repetitions they hear, as well as other clues, such as where speakers are gazing or pointing and the context.

The number and complexity of words produced at any age varies, but most children begin producing words between nine months and two years.[13] Recall from Chapter 5 that infants who performed best at perceiving the speech sounds of their native languages at seven months of age produced more complex words at two years old than other infants did.[14] Those seven-month-olds learned to perceive speech sounds largely through the motherese they heard. Motherese therefore appears to indirectly influence the words infants eventually produce and string together.

People generally have larger vocabularies for listening than for speaking, and this is also true for infants who are beginning to acquire words.[15] The motherese-primed ability to fish words from speech streams and to hear and produce the speech sounds that make up those words is, of course, essential to acquiring a vocabulary. But it does not stop there. The process *itself* of building a vocabulary contributes to the language acquisition that motherese started by helping infants refine their mental word banks:

For instance, a child with a lexicon containing only one word starting with /m/, *mommy*, can be rather lazy about how to represent and produce its phonological form (e.g., "muh"). Once more words that begin with /m/ are added (e.g., *milk* and *more*), phonological representations must become more complex in order to discriminate the words. The same could be true of semantic representations, causing naming errors as the child attempts to converge on more exact meanings.[16]

Parents also contribute by continuing the "naming game" that began when they started to interpret their babies' babbles. Studies on French- and English-speaking parents and their one- to two-year-old infants show that the extent to which parents incessantly label objects and encourage repetition of names is associated with their babies' vocabulary growth, as well as their ability to manipulate and categorize objects.[17] Firstborn children tend to learn nouns at a greater rate than younger siblings, perhaps because their parents have more time for the "naming game."[18] Infants who have higher rates of word acquisition tend to focus more on nouns, while more gradual learners seem to acquire a more balanced vocabulary of nouns and other types of words.[19] How quickly infants learn verbs depends on how often particular verbs occur when their parents talk to them.[20] And verbs are easier to learn if they exist at the end of sentences (e.g., "What's baby drinking?" rather than "Baby's drinking what?").

Although infants initially learn better when new words are in isolation, by the time they are eighteen months old, they understand familiar words more quickly if the words are heard in a phrase such as "Look at the ___" or "Where's the ___?"[21] An infant will gaze at a photograph of a dog more quickly if she hears "Where's the doggie?" rather than simply "doggie." Infants appear to benefit from the predictability of hearing words within such short familiar phrases, and parents intuitively seem to understand this, since they use these clipped phrases extensively when talking to their babies.[22] Karmiloff and Karmiloff-Smith say, "Even though infants do not understand everything in the speech that they hear during the first eighteen to twenty-four months, what parents actually say to them, and how they say it, can affect the nature of their subsequent word production. The variety of words used, the manner in which they are presented, and how often a child is addressed and drawn into speech-based interaction may all influence individual difference in word learning rates."[23]

The fine details also vary across cultures—Japanese and Korean children, for example, use more verbs than English-speaking infants.[24] Nevertheless, children in various cultures go through approximately the same sequence of stages as they acquire their languages.[25] Although individuals vary greatly, by the time they have reached two years of age, most toddlers can produce about fifty words, and these tend to be similar across cultures and languages: the names for parents ("Mommy," "Daddy") and other familiar people, animals ("doggie"), body functions ("wee-wee"), objects ("ball"), social routines ("bye-bye"), and imperatives ("up!").[26] As noted, babies everywhere usually utter their first words when they are between nine and twenty-four months of age. They then slowly add new words, but the rate soon increases. A two-year-old then learns to put two words together, especially for requests ("more juice"), which is about as far as language-trained apes get. Children usually begin to combine words grammatically when they can produce 150 to 200 words, and this is true even for developmentally delayed children, who begin producing words much later than is typical.[27]

GRAMMAR

As motherese aids with vocabulary-building, it also exposes babies to certain aspects of grammar. For example, baby talk helps toddlers acquire grammar through its intonations, which divide phrases into grammatically appropriate groupings. Although there is a huge amount of individual variation, children usually begin to produce grammatical speech between fourteen months to three years of age. By definition, speech becomes grammatical as infants conform to conventional rules for constructing words and combining them into phrases and sentences (syntax).[28] After babies learn how to produce them, certain words such as "more," "again," or "all gone" become useful "pivots" for creating two-word expressions. No one fails to understand when baby utters "more cookie," "all-gone juice," or "again peekaboo."[29] Word order emerges early in the development of English speakers as an especially important feature of syntax, and infants as young as seventeen months can use it to understand, for instance, the difference between "Big Bird tickling Cookie Monster" and "Cookie Monster tickling Big Bird."[30]

By the time they are two, toddlers can differentiate between verbs that take a direct object and those that do not (for example, "lifting" in "Big Bird is lifting Cookie Monster" versus "Big Bird is lifting with Cookie Monster"). Pinker suggests that once babies learn the meanings of relevant nouns, they can infer the meanings of such subtleties from the context in which they are heard:

> Upon hearing "The boy is patting the dog," for example, the child needs to know what the words "boy" and "dog" mean before he can even start a grammatical analysis of the sentence. Then, upon seeing the accompanying action (boy touching the dog's back), the child can use this real-word situation to make the formal linguistic analysis, mapping "the boy" to the subject noun phrase, and "patting the dog" to the verb phrase containing a direct object. . . . He can also derive from the linguistic context that "pat" is a transitive verb that must take a direct object. If the child later hears "the boy is running," he can again combine his knowledge of the meaning of "boy" with the transparent situation being referred to, to conclude that "run," a new word, is the verb phrase and means to move quickly.[31]

English speakers learn how to change the meaning of words by combining them with morphemes such as *s, ed, ing* and *un*, as well as how to use little words like "the," "a," "and," "on," "under," "in," and "out." Although the age at which children first produce morphemes varies, studies on American children indicate that they generally learn how to use morphemes in a particular order.[32]

Infants first produce *ing* to indicate continuing actions ("Daddy eating"). After that, they begin using the prepositions "in" and "on," followed by adding *s* to indicate a plural number. Using "a" and "the" with nouns occurs later, and the past tense (*ed*) appears even after that.[33] English-speaking mothers help their babies learn morphemes by repeating words back to them while adding the proper endings (such as *s* or *ing*).[34] As children use these morphemes more over time, they become better at combining and moving words within sentences.

Baby talk also influences the learning of grammar in languages such as French, Italian, Serbian, Polish, and Russian, where nouns are arbitrarily classified as masculine, feminine, or neuter. Because there is no logic to these assignments, learning them is difficult. Mothers help infants overcome this problem by using many diminutives (an example in English is "doggie" instead of "dog") that make these distinctions obvious.[35]

Although motherese helps infants learn words and grammar, it becomes less important as maturing children continue to build linguistic skills, and it wouldn't be effective if infants did not have nervous systems exquisitely tailored for language. Thus, nature and nurture are both important for language development.[36]

THE GREAT LANGUAGE-ORIGINS DEBATE

The association between exposure to baby talk and acquisition of language in modern infants raises the interesting possibility that the first language may have originated from some prehistoric form of motherese. In fact, the idea that the development of individuals (ontogeny) repeats the evolutionary development their species went through (phylogeny) has existed since at least 1866, when German zoologist Ernst Haeckel coined the phrase "ontogeny recapitulates phylogeny." Of course, today's biologists do not accept Haeckel's hypothesis in its literal form—after all, human embryos do not actually go through stages in which they are first fully formed fish, then amphibians, and so on.

Generally speaking, however, if a structure appeared earlier than another structure during evolution (which we know from the fossil record), it tends to precede the other during individual development. Human embryos, for instance, develop the deeper parts of their brains before the more recently evolved outer part (cerebral cortex).[37] All normal children also go through stages of being nonlinguistic and moving on four limbs before they become loquacious bipeds. And they teeter (delightfully) when they first walk, as our early ancestors probably moved when they were still refining their bipedalism. Furthermore, the burgeoning field of evolutionary developmental biology (known as evo-devo) now accepts a modified version of Haeckel's dictum, namely that "altering ontogeny formulates new phylogeny."[38] This newer version reflects that evo-devoists explore the roots of evolutionary change by identifying the precise developmental and genetic mechanisms that alter body shape and form in growing individuals.[39] At a broader level, the newly refined concept of "Haeckel Lite," as I like to call it, is consistent with the postural changes (from quadrupedalism to bipedalism) and increase in brain size that evolved in our ancestors and continue to be experienced during the development of individuals. It is also consistent with the idea that a prehistoric form of motherese may have been important for the origin of language, as it is for language acquisition of living infants.

The evolutionary importance of motherese has been questioned, however, because of an ongoing passionate debate about language origins. A particularly thorny issue is whether speech evolved out of earlier primate calls. One school of thought, often held by linguists, maintains that it did not.[40] An alternative view is that speech emerged incrementally and gradually from primate vocalizations under the influence of natural selection. Citing the well-known fact that evolutionary innovations are often built on earlier evolutionary changes, people who hold this view (and I count myself among them) see no reason why speech, and indeed the other aspects of language, could not have evolved from our early ancestors' apelike communications. These theorists tend to be receptive to the suggestion that motherese played an important role in language evolution.

As we have seen, some scientists concentrate on specific elements of language that are unique to humans (such as syntax), which can lead to a confusing partitioning of language into aspects that are shared with other animals and those that are not.[41] In a welcome departure, Jackendoff and Pinker investigated idioms and sayings, which led them to conclude that grammar resides in stored memories rather than in an inborn set of universal rules.[42] Rather than viewing syntax as a central core of language, they see it merely as a sophisticated accounting system that organizes meaningful relationships among words, phrases, and sentences.

From this "construction-based" view of language, Jackendoff and Pinker conclude, quite reasonably, that syntax could have evolved only after words emerged and the phonological abilities needed to produce them became refined.[43] This same sequence occurs as individuals develop, because infants' sensitivity to grammar (syntax) flourishes only after they have acquired a certain number of words and sharpened their phonological abilities through their exposure to motherese. It is therefore reasonable to assert (in keeping with the Haeckel Lite maxim) that a prehistoric form of baby talk facilitated the emergence of words and refinement of phonological abilities.

WHAT'S IN A NAME?

The first words hominin mothers spoke to their infants were probably much less complicated than modern words. In fact, some early words likely developed from signals with multiple meanings, like those used by birds, bees, and primates.[44] For example, the well-known calls of vervet monkeys show the potential complexity of primate vocalizations. In these monkeys, various alarm calls indicate the presence of leopards, snakes, and birds of prey and, upon hearing them, group members behave appropriately. Thus, the *chirp* that is produced upon spotting a leopard causes troop members on the ground to leap into trees, the *chutter* for a snake is met with visual scanning of the ground, and the *rraup* that is emitted upon spying an eagle causes vervets to look skyward and take cover.[45] Although these calls each signal a cry of fear, a warning, and a general reference for a specific dangerous predator, the monkeys are not able to use them as words in other situations.[46] Likewise, our ancestors would not have spoken their first true words until they extended their repertoire of vocalizations to include specific utterances that could be used under a variety of circumstances to indicate certain individuals, objects, or events.

So what did early words actually sound like? Various conjectures have been made over the years, many with fanciful monikers such as "yo-he-yo," "ding-dong," and "bow-wow." Proponents of "yo-he-yo" believe that words may have evolved from the sounds hominins made during collective physical activity. The "bow-wow" and "ding-dong" schools, on the other hand, suggest that hominins' first words imitated nature sounds—animal noises in the case of the former ("moo," "oink"), and other natural sounds for the latter ("crash," "boom," "clunk," "pow"). Such words are called onomatopoeic and, interestingly, mothers frequently use them when speaking to their babies. (Japanese mothers, in particular, favor onomatopoeic words.[47]) A favorite game that English-speaking mothers play with their infants is "What does a _____ say?" Mom asks the question and fills in the blank (e.g., "kitty"), and baby answers ("meow"). Experimental evidence also indicates that chimpanzees in the laboratory are better able to understand onomatopoeic words than other types.[48] Significantly, onomatopoeic words resemble real-world objects (sounds) rather than being abstract. Because of this, they are said to be iconic. (We will return to iconic communications when we discuss gesture and art in Chapter 8.)

Many scientists believe early words would have been names.[49] These could have referred to a variety of things or concepts, including foods, animals, predators, tools, places, weather conditions, dangers, discomforts, enemies, tribe members, and kinfolk.[50] In this context, recall that modern babies' initial vocabularies tend to be conceptually similar across the world's cultures and include the names for familiar people, animals, body functions, objects, social

routines, and imperatives. As the Haeckel Lite theory would predict, to some extent modern vocabulary development in individual children may parallel the evolution of that in our ancestors.

Prehistorian Steven Mithen suggests that some of the first vocal expressions learned by infants may have been "yuk" or other
closely related sounds of disgust such as "eeeurrr," which are found
in all cultures today and, when made by parents, are usually accompanied by wrinkled noses and pulled-down corners of the
mouths.[51] According to Mithen, such words would have been
useful because modern (and presumably prehistoric) babies do
not react with disgust to "bodily excretions, decaying food, and
certain types of living creatures, notably maggots" until they are
between two and five years of age. Mothers, then, would likely
have uttered sounds of disgust to dissuade their babies from in-
gesting dangerous substances. Mithen adds that his idea has at least
one critic, namely his wife, who believes that *yumyumyumyum
yummm*, the sound parents make when trying to persuade babies
to eat some foods, may have had evolutionary priority over *yuk*.

Research on the English word "mama" is particularly relevant
to discussions on the origins of language.[52] Linguist Peter Mac-
Neilage points out that "mama" consists of two syllables that begin with a consonant, end with a vowel, and are generated by
moving the lower jaw while keeping the tongue still.[53] (You can
easily verify this.) MacNeilage thinks that such words probably
typified the earliest speech. Babies begin producing the word
"mama" at around two months, usually as part of a cry, and some
infants seem satisfied if a caregiver responds by paying attention to
them, while others also want to be picked up.[54] Parents, of course,
interpret babbles that approximate "mama" to mean "mother" and
eventually succeed in conveying this word to their infants. By the
time they are six months old, infants understand that the word
"mama" specifically refers to *their* mom rather than to any
woman. This suggests that, thanks to the repetitions of motherese,
they have begun to build a vocabulary of sounds that represent
significant people.[55] I love the thought that one of the earliest
words invented and shared by our ancestors might have been the
equivalent of "mama." After all, wouldn't babies then, as now, have

been inclined to put a name to the face that provided them with their earliest experiences of warmth, safety, and love?

It is easy to envision early hominin mothers humming and cooing to their infants while engaging them in playful vocal exchanges. That such games would have entailed mimicry and repetition is not surprising, and increasingly big-brained early moms would eventually have begun to attach meanings to their infants' repeated syllables. For their part, infants would in due course begin understanding the meanings of these new words that, in some cases, they had unwittingly helped to invent. Over enough time and under the influence of both natural and cultural selection, these interactions would have had a significant impact on both the development of infants and our species as a whole. In other words, just as motherese would have increased the reproductive fitness of hominin mothers and infants (as we discussed in Chapter 2), hominins who acquired words and eventually protolanguage would have had an edge over those who did not.

But after words were invented, what was the next step? How would protolanguage have emerged?[56] Everyone seems to agree that at some point after our ancestors acquired a critical mass of shared words, they began combining them into simple utterances. Language likely developed when our ancestors began to realize they could produce deliberate, symbolic vocalizations, string them together, and be understood by others, as modern babies realize during their second year. In fact, a precedent for inventive insight among higher primates illustrates how such a thing might have happened.

Japan is known for its clever, inventive monkeys (*Macaca fuscata*). These macaques are famous for having invented new practices that became widespread across monkey society and, remarkably, have been passed down through the generations since the 1960s. In contemplating how words may have initially emerged and become conventionalized in our ancestors, we can look at the way Japanese macaques developed new customs. Some of the snow monkeys from Honshu Island, for example, have become accustomed to keeping warm in the winter by soaking in natural hot springs, a habit that was introduced to them by a female monkey who tried it after observing human bathers. On another island, a female monkey named Imo developed the practice of rinsing the gritty sand off sweet potatoes, and later began seasoning the potatoes with salt by dipping them in seawater.[57] Within nine years, all of the monkeys in her troop except for the very youngest and oldest had learned to wash and salt their sweet potatoes like Imo. Impressively, the descendants of those monkeys still do so today.

Imo's second invention was even more startling. On the island of Koshima, where Imo lived, primatologists regularly threw wheat on the sand for the macaques. When the monkeys attempted to gather the wheat, however, they ended up with an unappetizing sandy mixture. One day, Imo scooped up a handful of the sandy wheat and stood next to the water. She looked at her hand, and then at the water. In an apparent moment of inspiration, Imo threw the handful of the mixture onto the sea. The sand sunk, and she skimmed off and ate the floating grains of clean wheat. Had Imo experienced a logical insight? Although we can never be certain, one thing is for sure: after Imo invented this technique, the rest of the troop adopted it and has handed it down through the generations ever since. Led by another female, the Koshima monkeys have also learned to wade, dive, and swim in the sea. The inventiveness of Japanese macaques continues today: they are still inventing food-preparation techniques, including a recent one for washing grass roots by rolling them on flat rocks.[58]

Particularly relevant to the origin of protolanguage in our ancestors are the documented stages that *preceded* the appearance of widespread new behaviors in Japanese macaque society.[59] In the first "Period of Individual Propagation," the offspring of the discoverers taught the novel behaviors to their peers (in play groups), and those youngsters then taught the behaviors to their mothers and older siblings. Adult males were the last to acquire new practices. After behaviors became established across social groups, a second "Period of Pre-cultural Propagation" began, in which new infants learned the customs directly from their mothers. Thus, the practice began to be passed from generation to generation. It is especially significant that the monkeys' novel behaviors were introduced by females and eventually became established across generations through mother–infant communication.

Japanese macaques are not the only primates whose females seem to be the more inventive sex. Chimpanzees that live in the Taï Forest of Africa's Ivory Coast are well-known for using rocks to crack open nuts, and females engage in this activity more than males.[60] There also are anecdotal reports of mothers actually *teaching* their infants how to use hammer stones to open nuts. Primatologists were recently astonished to learn that chimpanzees living in Senegal bite the tips of sticks and use them to spear bush babies (another kind of primate), which they eat. This kind of tool preparation combined with hunting was, until now, thought to be unique to humans. Again, females and immature chimpanzees appear to practice this new custom more frequently than adult males.[61]

Apart from studying modern primates, how else can we reasonably speculate about the emergence of vocabulary during hominin evolution? Jinyun Ke and colleagues from City University of Hong Kong used simulation and mathematical models to explore the evolutionary emergence of vocabulary, and their results are striking.[62] They began with the assumption that there was a stage during which early hominins became aware that vocalizations could be symbolic. In other words, they had acquired a prehistoric "naming insight." Initially, individuals would have produced arbitrary "names" that covered broad categories, similar to the alarm calls of vervet monkeys. But how, the authors wondered, did some of these names eventually come to be accepted by all group members?

Ke's "hybrid model" includes calculations about how vocabulary could have spread horizontally across a population of early ancestors and then been transmitted vertically from generation to generation. The model assumes that, at first, our ancestors would have individually produced vocalizations that referred to objects. Imitation would have been a catalyst for forming a common vocabulary, as individuals interacted and gave up their own arbitrary names in favor of those used by others. By doing so, "the local, individual actions of many speakers, hearers, and acquirers of language across time and space conspire to produce non-local,

universal patterns of variation."[63] One can therefore think of
these early hominins as having responded to a kind of linguistic
peer pressure that eventually caused names to be adopted across
groups. Certainly this happens today: consider the new vocabulary spread throughout large regions today by teenagers.

An important aspect of Ke's model is the observation that
words are transmitted from one generation to the next because
children learn them from their parents. Over time, such transmission favors words that are consistent and clear rather than ones
that are confusing or ambiguous—a sort of "survival of the fittest"
for words. This would have happened in the past, too, resulting in
cultural (rather than genetic) selection of vocabularies. The authors conclude that the combined effect of the horizontal spread
of words together with vertical cultural selection resulted in high-
speed vocabulary evolution. Interestingly, Ke's simulation suggests
that conventional vocabularies emerged in groups that were neither
too small nor too large, possibly ten to fifteen individuals. The
model also suggests that, after a long period of oscillation, there
was a tendency for a global vocabulary to emerge relatively suddenly. Of course, once a vocabulary became fixed across a small
group and then transmitted from generation to generation,
a population explosion would have sufficed to make it widespread.

Think back now to the emergence of sweet-potato salting
and wheat washing in Japanese macaques. The horizontal and vertical transmission of these cultural inventions proceeded *exactly*
according to Ke's model for vocabulary emergence in our ancestors. In my opinion, so too would have the horizontal and vertical
transmission of elementary aspects of grammar in hominins, such
as conventions for forming two-word phrases. In fact, the transmission of cultural inventions in Japanese macaques and Ke's
model are both compatible with the suggestion that developmental changes in early hominins sparked the emergence of the first
domino—motherese—in an unfolding chain of events that eventually led to the evolution of language.

Although it is essential to realize that mothers and infants
contributed significantly to the evolution of language, they may
also have influenced numerous other key aspects of communica-

tion. For example, people communicate not just with words, but also with tone of voice, which I believe was the predecessor to language's sister, music. In addition, we do not just talk, we simultaneously gesture, which may have been an impetus for the evolution of the temporal and other arts. We will explore these topics in the next two chapters.

CHAPTER 7

SHE SHALL HAVE MUSIC

BANBURY CROSS

Ride a Cock-horse

To Banbury Cross,

To see an old lady

Upon a white horse,

Rings on her fingers,

And bells on her toes,

She shall have music

Wherever she goes.

—Mother Goose

L IKE THOSE OF LANGUAGE, the origins of music have been
hotly debated since at least the nineteenth century. What is
the evolutionary purpose of music? How did it develop? Some
scholars believe music is a useless by-product of the neurological machinery that evolved for language. Music, they claim, is
what Stephen Jay Gould called a spandrel.[1] In architecture,
spandrels—decorated spaces between the arches and molds
of buildings—have no architectural purpose. Evolutionary biologists have co-opted this term to label features that are

by-products or side effects of other traits rather than direct results of natural selection. One of the best-known spandrelists is Steven Pinker, who believes music is an "exquisite confection" that borrowed "some of its mental machinery from language—in particular prosody." According to Pinker, music is therefore nothing more than "auditory cheesecake."[2]

On the other hand, Charles Darwin rejected the idea that music evolved from language. Taking a cue from birdsong and gibbons, Darwin argued that speech developed from musical notes and rhythms that had evolved in our primate ancestors "for the sake of charming the opposite sex."[3] Although numerous contemporary scientists agree with Darwin's music-first hypothesis, many disagree that music evolved for attracting mates.[4]

Before we dissect music's evolutionary purpose and its relationship with motherese, we should define music itself. Scholars generally agree that music, like language, is found in all cultures and, just as different cultures have distinct languages, they have different musical traditions. In most societies, music is used and made by many people. In Western society, on the other hand, such activities are frequently relegated to specialists. In addition to who makes music, the definition of music varies with culture. Some traditional societies do not even have a word for music in the sense that we do. The Blackfoot word *saapup*, for example, means something like dancing, singing, and ceremony all rolled into one.[5]

The quest to identify universals in music is fraught with problems. Nevertheless, ethnomusicologist Bruno Nettl suggests that all societies have vocal music and some sort of musical instruments, although these can be as simple as a fallen tree trunk to drum on. They all have some music that contains a meter or pulse, as well as music that uses only three or four pitches. Music is universally used to mark important events and to accompany dance. It is important in ritual and other contexts associated with the supernatural, and it is used to alter the consciousness of individuals or the ambience of social gatherings. From the perspective of the role that lullabies may have had for music evolution (discussed later in this chapter), it is interesting that Nettl believes the world's simplest and most widespread music consists of songs with short repeated phrases, which are often used in children's games in societies with more complex music.[6]

Given the universality of music, it is not surprising that it inspires so many different definitions. For some, "music, like speech, consists of complex sound patterns that vary over time."[7] Pitch and rhythmic structure, which convey and induce emotions in both music and motherese, have been identified as the two main dimensions of music.[8] Music has also been conceptualized more broadly "as a behavioral and motivational capacity: what is done to sounds and pulses when they are 'musified'—made into music."[9] Prehistorian Steven Mithen (Reading University, England) goes even further, noting that one of the most striking features of music is "bodily entrainment," in which people participate by tapping their fingers and toes and sometimes moving their whole bodies. It is artificial, he believes, to separate rhythmic and melodic sound (music) from rhythmic and melodic movement (dance). Mithen uses the word "music" "to encompass both sound and movement."[10] This comes closer to the way music is conceptualized in many traditional societies.[11]

The essence of music has also been defined more formally. Just as languages have inventories of sounds, boundaries of sound categories (e.g., words, phrases), and rules for combining them (syntax), musical systems have inventories of pitches (notes) organized into scales and a syntax for arranging them into acceptable sequences.[12] Because of our grasp of syntax, a sentence such as "Jane see Spot" feels wrong. Similarly, the distinctive four-note "short-short-short-long" opening to Beethoven's Fifth Symphony (*da da da dum*) would feel wretched without that last *dum*. Musical syntax regulates the formation of harmonic structure (chords, chord progressions, and keys) and sets up our expectations for what should come next. Like poor spoken grammar, poor harmonic grammar is implicitly grasped by listeners, which is why one does not have to be a trained musician to detect sour notes.

Aniruddh Patel of the Neurosciences Institute in San Diego has studied how the brain processes the syntax of both language and music, and he has come to the significant conclusion that the two are much more closely related than previously believed.[13]

Patel notes that listeners mentally connect each element of incoming speech (words) or music (notes) to other elements as the speech or music unfolds. He gives the following example of a sentence in which listeners automatically connect "sent" to "reporter": "The reporter whom the photographer sent to the editor hoped for a story." Such syntactic processing is rapid, unconscious, and tricky, and it becomes more complicated with increased distances between pairs of elements that require connecting. The same is true for processing the tones of incoming music, which results in listeners experiencing patterns of tension and resolution as the music develops. It is fascinating that linguistic and musical experiences share certain steps that occur early during syntactic processing and that these steps take place in a part of the frontal lobes that includes Broca's "speech" area.[14] To further investigate the relationship between language and music, Patel's team measured the duration and pitches of vowels in spoken English and French and compared them to the melodies and rhythms of instrumental music from the two cultures. They found that the rhythms and cadence of the spoken languages are reflected in their national music.[15]

THE MEANING OF MUSIC

Despite their different views regarding what, exactly, music is, researchers agree unanimously about what separates it from language. Although both systems are composed of discrete units—tones for music and words for language—only words are symbolic. Because words refer to arbitrary objects and concepts, they can be strung together to convey precise information. This is not the case for tones, which can be strung together only to express and evoke emotions by creating patterns of tension and resolution.

And musical tones certainly do evoke emotions. At one level, this is obvious because people usually agree on the kinds of music that make them feel sad, happy, frightened, and so on. Recall, for instance, the alarming, high-pitched, screechy music that accompanies the shower scene in the movie *Psycho*. The emotional responses to music can be so strong that some people experience shivers when they listen to music that is particularly pleasurable or moving. The emotions that music evokes are associated with the distances between notes as the music unfolds and, in particular, where in the pattern the halfnotes (semitones) occur. For example, musicologists have observed that positive emotions such as love, peacefulness, and joy are often expressed in Western music by major scales that have their first half-steps between the third and fourth notes, while negative feelings, such as sorrow and fear, may be evoked by minor scales, in which semitones occur between the second and third notes.[16]

Neuroscientists are just beginning to unravel how music is experienced in the brain.[17] Robert Zatorre of McGill University has shown that a part of the right frontal lobe is important for perceiving pitch and remembering music, in contrast to language, which is processed largely on the left side of the brain.[18] In fact, the right side of the brain is generally considered to be the "musical" half and is more involved with understanding and expressing emotions than the left.[19] (However, the left hemisphere processes the more analytical aspects of music, especially for trained musicians.) The right side of the brain also excels at visual, spatial, and mental imagery, and has an edge over the left hemisphere in recognizing faces and understanding jokes and metaphors. Unlike the verbal and analytical left side of the brain, which grapples with little sequential details, the right side is intuitive and grasps information in a more global manner.

Most significant, it is the right side of the brain that provides tone of voice or prosody when we speak. Like music, the meaning of prosody is emotional. Tone of voice adds color and emotional overtone to the meaning of spoken words. While the left sides of our brains rapidly interpret the literal meaning of unfolding streams of words, our right hemispheres decipher what speakers *really* mean by interpreting their tone of voice. Like body language, tone of voice sometimes speaks louder than words. For example, Anthony Hopkins's tone of voice adds deliciously to his frightening character, Hannibal Lecter, in the film *The Silence of the Lambs*.

Because tone of voice provides emotional nuances for speech, it is an integral musical component of language today. But how functionally important were prosodic vocalizations and music during hominin evolution? Were they simply spandrels? Or were prosodic vocalizations, singing, and other ways of making music targets of natural selection in their own right?

If we want to unravel our musical origins, there is much to learn from our nearest relatives, the nonhuman primates. Singing is unusual among primates, occurring in only two groups of prosimians (tarsiers and indris), one New World monkey (titi monkeys), and one ape (gibbons).[20] In all four of these groups, monogamous pairs of males and females live in and defend territories, and defend (more or less) exclusive sexual access to each other. Monogamy is very rare in primates, but when it does occur, it includes singing.

The most exquisite singing of any primate, indeed of any land mammal, occurs in the dozen or so species of gibbons, which are known as lesser apes because they are much smaller than the three great apes (orangutans, gorillas, and chimpanzees). With luxurious coats, bare black faces shaped like three-leaf clovers, and large eyes, gibbons are, in a word, beautiful.[21] They are also skilled gymnasts that glide gracefully, hand-over-hand, through the trees. Although males of most gibbon species sing long solos right before dawn, females later join them in duets that sound highly appealing to humans. The songs of gibbons have one central message: "This is my territory, my mate. I'm healthy and strong—as you can tell from my wonderful voice—so keep away." Because they are bonded to one partner, paired gibbons do not sing to attract mates (however, non-mated males seem to). Instead, their singing is thought to help strengthen the pair bond.[22]

Gibbons are born knowing their songs rather than having to learn them from others the way songbirds and people do.[23] But new evidence suggests that gibbons have greater voluntary control over their vocalizations than previously believed.[24] Indeed, acquiring proper songs may entail at least a certain amount of learning in gibbons, illustrated by the fact that "infants often squeal during a mother's great call, and juveniles may join duets with imperfect, sex-appropriate sounds."[25]

Although gibbons are the only apes that sing, each of the three great apes uses vocalizations to control the distance between individuals or groups. All of these distance-controlling calls resemble elements of gibbon song, especially the female great calls (song phrases consisting of a series of ascending notes). For example, orangutan males produce long calls that are often accompanied by branch shaking, and these can be heard at great distances—"keep away, males; come hither, females." Male gorillas engage in hoot series that may end with beating their chests, breaking branches, and thrashing through foliage. The hoots begin softly, and build up in intensity during a call, which can be heard up to a mile away.[26] Two or more males have even been observed vocalizing in a manner that appears to foreshadow singing: "One began hooting only to trail away to nothing before trying again. Then another joined in, and a third one. Their clear hu-hu rose and fell as each stopped and started independently. But when one reached a climax and beat his chest the others followed. Then they usually settled down for a few minutes before repeating the whole procedure."[27]

Our closest relatives, chimpanzees, also use long-distance vocalizations called pant-hoots. Males, who produce them much more frequently than females, utter a series of loud calls that rise and fall in pitch and often end with a scream. Chimpanzees recognize individuals from their pant-hoots, which are used while capturing prey or to answer others, announce sources of food, and threaten unknown chimpanzees. Different groups of chimpanzees have distinct pant-hoots (similar to human accents), which suggests that chimpanzees tailor their pant-hoots to be different from those of their neighbors.[28] In other words, pant-hoots are not just genetically determined, but also require vocal learning. Although chimpanzees do not sing the way gibbons do, Jane Goodall heard pant-hoot "choruses" during the night, passing back and forth between chimpanzees nesting in different trees.[29] She also noted that males sometimes engaged in drumming their hands and feet on large trees, often while pant-hooting.

Because of the similarities shared by gibbon songs and the calls of great apes, primatologist Thomas Geissmann suggests that singing and calling in all of the apes evolved from a common ancestral form. These ancestral calls would have contained pure

FIGURE 7.1. An approximately thirteen-year-old chimpanzee named Sherry drums on a can at Gombe in 1974. *Photo by Curt Busse.*

notes, set phrases, and an accelerated rhythm that slowed down near the end. Above all, the calls would have been loud enough to communicate over long distances. Geissmann also notes that "the human voice has often been identified as the most ancestral instrument used in music," and speculates that our ancestors' loud calls may have been the source from which human singing and, ultimately, music emerged.[30]

Geissmann points out several characteristics of human music that are *not* found in ape calls or singing, which suggests that these traits emerged after hominins diverged from their apelike ancestors. Apes cannot keep time, so a sense of steady rhythm probably evolved more recently than loud calls or songs. The learning of phrases and ability to improvise would also have developed more recently. Our early ancestors' music most likely would have been used "to display and possibly reinforce the unity of a social group toward other groups. In humans, this function is still evident today whenever groups of people, be they united by political, religious, age, or other factors, define themselves by their music. National hymns, military music, battle songs of fans and cheerleaders encouraging their favorite sports teams, or the strict musical preferences of youth gangs may serve as examples of this phenomenon, whose origin may go back to the very beginning of human evolution."[31]

A wide variety of animals, including primates, produce another kind of call that also controls distance between individuals. This vocalization is so vital that it must have been directly and pervasively targeted by natural selection. In fact, it may be at the root of all other distance-regulating calls, including those that eventually evolved into human music. I am referring to the relatively short-distance contact calls that infants and mothers produce when they become separated. Typically, lost infants give distinctive calls that are recognized by mothers as coming from *their* offspring. House-mice pups produce ultrasonic calls, lost infant opossums make sneezing sounds, baby horseshoe bats produce broadband calls, birds that have fallen from their nests emit isolation peeps, and dolphin infants produce isolation whistles when they swim too far from their mothers.

Contact calls are also common among primates. To name just a few of the groups in which they occur, isolated golden potto infants produce *clicks* and *tsics*, strayed baby aye-ayes emit *eeeps* or *creees*, lost infant squirrel monkeys produce easily localized and individually recognizable isolation *peeps*, and separated chimpanzee infants utter *hoos*. Whatever their species, mothers usually respond by answering their infants' contact calls with special ones of their own—and then go find their infants. Obviously, mothers and infants who do this would have a survival edge over those who do not.

In fact, John Newman of the National Institutes of Health and his colleagues suggest that the brain circuits used for mother-infant communication developed early during the evolution of mammals and that these same circuits, which help process emotions, become active when human mothers hear their infants cry.[32] The definition of motherese can thus be broadened to include the special infant-directed vocalizations of mammals.[33] Because the parts of the brain used for mother-infant vocalizations

are also important for speech, Newman believes that "spoken language may have arisen from hominin females vocalizing to their
infants."[34] Other scientists have also drawn specific parallels between the contact calls of primate mothers and human motherese. For example, the use of pitch and contour in the caregiver
calls of mother squirrel monkeys has been compared with the use
of motherese by humans.[35]

So how can gibbon singing, loud calls of apes, and contact
calls of mothers and infants help to shed light on the evolution of
human music? The research of Nobuo Masataka from the Primate
Research Center at Kyoto University in Japan addresses this issue.
Masataka is an expert on communication in children and nonhuman primates and is known internationally for his research on
music, evolution, language, motherese, and sign language. He and
his colleagues have shown that when Japanese macaque mothers
try to attract their infants' attention, they coo in a way that appears to have parallels with the prosody of human motherese.[36]

Although humans gradually produce speechlike sounds during their first year of life, Masataka observes that they do so in two
stages, the first of which entails the production of vowel-like coos
between six and eight weeks of age. Babies then learn to produce
speech sounds through social turn-taking with caregivers, and the
quality of adult vocal responses affects the vocalizations of developing infants. According to Masataka, the same process takes place
in Japanese macaques, who utter coos similar to those of humans,
not just as contact calls between mothers and infants but also (using a slightly different coo) to strengthen relationships and maintain contact with other group members. He also notes that
Japanese macaques' coos are used to maintain group cohesion in
place of physical contact, which has interesting parallels with the
idea that the voices of early hominin mothers were used in place
of their cradling arms.

In vocal exchanges between adult Japanese macaques, "the
temporal patterns of occurrence . . . between two consecutive
coos . . . were similar to those obtained in human mother-infant
dyads; after a monkey cooed spontaneously, it remained silent for
a short interval, and if no response was heard from the other

FIGURE 7.2. Nobuo Masataka is shown here in front of an open enclosure of Japanese macaques at the Primate Research Institute, Kyoto University. *Photo courtesy of Chisako Oyakawa.*

monkeys, then the monkey would often coo again to address the other group members."[37]

This observation makes clear that infant-directed calls by mothers (and their infants' responses) may have been important for the evolution of more general contact calls between group members. In other words, vocal communication between group members may have emerged from a kind of monkey motherese, similar to the evolution of human speech.

The second stage Masataka recognizes in the development of human speech is babbling, which begins before eight months. (We saw in Chapter 5 that human babies bridge Masataka's two stages by developing the ability to produce complex melodics in their cries.) Although nonhuman primates cannot babble the way human infants do, Masataka observes that their long-distance calls resemble the structure of babies' early babbles. He sees parallels between early babbling in humans and the pure tones, set phrases, and accelerated rhythms Geissmann described in ape loud calls and gibbon songs. Although Geissmann sees loud calls as having given rise to singing and, ultimately, to human music, Masataka goes a step further by suggesting that gibbonlike songs also gave rise to speech, perhaps through a process in which duets split into completely independent solos that over time developed into speech.

Darwin also focused on gibbon songs, concluding that human speech was probably derived from musical notes and rhythms that had evolved in our primate ancestors. Tecumseh Fitch of the University of St. Andrews, Scotland, notes that Darwin's idea is convincing because "the many similarities between music and language mean that, as an evolutionary intermediate, music really would be halfway to language and would provide a suitable intermediate scaffold for the evolution of intentionally meaningful speech."[38] Darwin also believed that music evolved for "charming the opposite sex." Although this idea has supporters, Fitch disagrees, citing the very early development of music perception and singing in human babies and the universal use of songs between mothers and infants.[39] Unlike the musical abilities of infants, sexually selected traits usually do not appear until sexual maturity. Fitch concludes that "the childcare hypothesis represents the account of the adaptive function of music currently most firmly grounded in data."[40]

The child-care hypothesis, of course, says music evolved from mother–infant communication. Along these lines, Inge Cordes of the University of Bremen, Germany, studied the relationship between melodies found universally in motherese and melodies in songs from sixty countries.[41] The melodies in motherese included rising ones that catch infants' attention, softly falling tunes that soothe infants, gently inclined bell-shaped ones used for approval, and steeply falling bell-shaped melodies that discourage behaviors. The four types of songs Cordes studied were lullabies, songs to arouse attention, praise songs (that approve desired behavior), and warriors' songs. She discovered a similarity between the melodic pattern of each song type and the pattern that expresses similar emotions in motherese. Songs to arouse attention have melodies that rise steeply, praise songs reveal gently inclined bell-shaped melodies, and warriors' songs have mostly steeply falling melodies. Although lullaby melodies turned out to be somewhat more varied, these findings show that melodies of motherese associated with certain emotions are found in both infant-directed and adult-directed songs.

But that's not all. Cordes also measured the average durations of the songs and compared them to the durations of different animal calls. Across a wide range of animals, attractive distance-shortening calls have a long rise and are stretched out, whereas calls that repel are noisy and short, and rise quickly.[42] Again, she found a clear correspondence: praise songs and lullabies reflect the stretched melodies of attractive animal calls, while warriors' songs have quickly rising melodies. Cordes also observed that songs to arouse attention have extremely long melodies similar to animal alarm calls. These results led Cordes to agree with Darwin's idea that music—especially singing—evolved from earlier forms of primate vocal communication.

The most basic function of music, of course, is to express emotions, as illustrated by neuroscientist Jaak Panksepp of Washington State University and Günther Bernatzky from the University of Salzburg in Austria. Together they have studied the types of music that cause many people to shiver or have chills, such as bittersweet songs of unrequited love and longing, or music that expresses patriotic pride, such as "The Battle Hymn of the Republic."[43] Females tend to react with shivers more than males, often in response to new or unexpected crescendos or harmonies. Sad music can also evoke the chill response, and Panksepp and Bernatzky's observations about this are particularly poignant in light of both contact calls and motherese:

> In our estimation, a high pitched sustained crescendo, a sustained note of grief sung by a soprano or played on a violin . . . seems to be an ideal stimulus for evoking chills. A solo instrument, like a trumpet or cello, emerging suddenly from a softer orchestral background is especially evocative. Accordingly, we have entertained the possibility that chills arise substantially from feelings triggered by sad music that contains acoustic properties similar to the separation call of young animals, the

primal cry of despair to signal caretakers to exhibit social care and attention. Perhaps musically evoked chills represent a natural resonance of our brain separation-distress systems which helps mediate the emotional impact of social loss. In part, musically induced chills may derive their affective impact from primitive homeostatic thermal responses, aroused by the perception of separation, that provided motivational urgency for social-reunion responses. In other words, when we are lost, we feel cold, not simply physically but also perhaps neuro-symbolically as a consequence of the social loss.[44]

MUSIC IN THE CRIB

Babies everywhere are extraordinarily musical.[45] Experiments show that even tiny infants can detect the smallest differences in pitch and timing that are musically meaningful across cultures.[46] Like adults, babies can remember melodies and can recognize them after they have been transposed to a different key or played at another tempo.[47] By the age of two months, infants also prefer consonant musical sequences to dissonant ones. These "rudiments of music listening," concludes Sandra Trehub, "are gifts of nature rather than products of culture." Sensitivity to culturally specific aspects of tonal and harmonic structure, on the other hand, emerges much later, between the ages of five and seven years.[48]

It is not surprising, then, that lullabies and playsongs are sung to infants throughout the world (as we discussed in Chapter 2). So how does infant-directed music differ from other music? Laurel Trainor and her colleagues at the Auditory Development Laboratory at McMaster University decided to find out by recording mothers singing directly to their infants and, at other times, singing the same songs when their infants were absent.[49] The performances turned out to be very different and everyone, including the babies, could feel the difference. Lullabies that were sung directly to babies generally had higher pitch, slower tempo, and a more regular rhythm—qualities associated with affection, tenderness, happiness, and arousal. The infants' presence also seemed to alter the mothers' emotional states, which, in turn, affected their voices, and adult raters observed that only the infant-directed songs sounded as if the mothers were smiling as they sang.[50] In fact, infants might have preferred the live as opposed to recorded songs because "soothing singing likely calms mothers as well as infants."[51]

Because babies generally prefer to listen to higher rather than lower speaking voices, Trainor thinks the unconscious preference for speaking in a higher pitch to infants may be biologically rooted. This is consistent with the association of low-pitched sounds with aggressive and hostile displays in animals (as in *growl*) and high-pitched sounds with more friendly situations (though higher pitched sounds are also associated with fright). But other factors can trump this preference. For example, features of comforting or loving motherese, such as lowered pitch, slow tempo, and downward-gliding pitches, are also found in lullabies, which are sung more slowly and with a more regular rhythm than playsongs.[52] Though infant-directed playsongs have more rousing features, lullabies have features that calm babies and induce sleep.

Trainor's team also discovered that mothers exaggerated cues about the musical structure of songs when singing directly to infants, unconsciously singing in a way that helps develop babies' musical perception, similar to how motherese helps infants learn to break down speech streams. Motherese and infant-directed songs therefore appear to have multiple and parallel functions: to attract babies' attention, to communicate and control emotions, and to teach infants about patterns of sound. As such, lullabies and playsongs are a vehicle for a kind of motherese that focuses on music (musicese?) rather than speech. And it's not just mothers who sing this way. The presence of a baby triggers similar effusive singing in fathers, nannies, other bystanders, and even preschoolers.[53]

Babies everywhere respond favorably to maternal singing, and infant-directed songs capture their attention even better than motherese.[54] Partly for this reason, Elena Longhi and Annette Karmiloff-Smith of University College London argue that singing—rather than verbal motherese—may have set the prehistoric stage for early mother-infant interactions.[55] Songs also have a more regular musical structure than speech, making them easier to break down into smaller units, although the lyrics of lullabies (and playsongs) are completely irrelevant from the babies' perspective:

> One crucial difference between motherese and song is that motherese tends to stress meaning alongside social interaction, whereas the actual semantic content of songs is often completely irrelevant. What is crucial about songs is the rhythmic and segmental characteristic of the vocal message, and this may make it primordial over early linguistic interaction.[56]

Longhi and Karmiloff-Smith emphasize that, instead of listening passively to their mothers' songs, infants react with emotional responses such as smiling and cooing, and coordinate their movements with the timing of the song.[57] In turn, moms mark the boundaries between the phrases of their songs by shaking their heads, patting their babies with tightly synchronized movements, and so on. Like motherese, infant-directed music is actually a kind of multimodal dialogue—a duet.[58]

Bouncing babies in time to music also actually helps them develop musical sensitivity.[59] Trainor noticed that parents couldn't refrain from bouncing, rocking, or playing with their babies' feet when they sang to them, and she wondered whether moving the babies was somehow important for their development. She and her colleague Jessica Phillips-Silver had seven-month-olds sit on their mothers' laps while listening to a rhythm with no accented beats played on a snare drum. The mothers created an accented rhythm, however, by bouncing their babies. Half of the mothers were instructed to bounce on every second beat, and half on every third beat. Infants were then tested to see which of the two accented rhythms they preferred to listen to by measuring how long they looked at the loudspeakers playing the two patterns. Even though they had *heard* exactly the same unaccented rhythm, the infants preferred to listen to the rhythm that matched how they had been bounced. Trainor and Phillips-Silver concluded that there is a strong, early connection between the processing of body movements and rhythms that are heard simultaneously, and that it is critical for human musical development.[60]

Both infant-directed songs and motherese are crucial for babies to understand and express emotions. Although babies' first words thrill parents, the equally impressive achievement of their growing ability to process and express emotions is more subtle and perhaps not as deeply appreciated. In fact, learning to decipher speakers' attitudes and emotions from nonlinguistic cues (speaking rate, voice quality, and swings in pitch) is extremely important. Tone of voice contributes greatly to mature listeners' interpretations of the words and phrases of speech.

In the face of conflicting cues, adults assess speakers' feelings by how they speak, not what they say.[61] On the other hand, children younger than eight years of age judge speakers' feelings by speech content rather than tone of voice. Although children as young as seven know when speakers say happy things in sad voices (or vice versa), they have not yet developed an adultlike skill for interpreting tone of voice when assessing speakers' emotional states. This shows that, despite infants' intense exposure to prosody through motherese, it takes years for them to learn to process the literal meaning of speakers' words in light of subtle nonlinguistic cues. (In a similar vein, young children have difficulty understanding irony and sarcasm, which require integrating nonlinguistic cues with opposing literal meanings.[62]) That accurately processing the emotional undertones of language is such a hard-won yet pervasive skill suggests that it was important during human evolution.

WHENCE MUSIC?

Evidence from animal contact calls and infant-directed vocalizations suggests that the prosodic vocalizations from which music arose were indeed crucial for survival during hominin evolution. Falling out of the nest would have been deadly unless moms heard their infants' distress calls and retrieved them. Prehistoric mothers who allowed their tiny infants to cry incessantly as they picked berries or dug tubers would have made them attractive targets for hungry predators.

Our capacity for processing tone of voice helps us attribute mental states to others, an ability some scientists think was intensely selected for as our ancestors evolved.[63] I also believe that the musical quality of normal human speech is still important for individual survival today. Tone of voice reveals speakers' fluctuating emotional states and provides listeners with unconscious clues. From speakers' tones, we perceive their emotional involvement in the conversation and the likelihood that they are being truthful. We may even pick up occasional warnings that speakers are going to do harm. An odd or inappropriate tone of voice is off-putting to listeners, and heeding that uneasy feeling can be very adaptive, allowing the listener to avoid potentially dangerous people and situations.[64]

All normal people develop speech, but not everyone plays a musical instrument or composes symphonies. For this reason, many scientists (including me at one point), have suggested that language was a direct target of natural selection but that music was not.[65] However, I no longer believe this to be true. With input from the right side of the brain, people make music every time they speak. If tone of voice is the music in speech, then the lyrics of songs are the speech in music. (Motherese has been described as both "spoken music" and "musical speech."[66]) I now believe that music and language evolved in lockstep both with each other and with the enlargement and rewiring of both sides of the brain over millions of years.[67] This is not to say that Beethoven's piano concertos are the *direct* result of natural selection any more than recitations of Shakespeare's sonnets or Walt Whitman's poems are. But one thing seems clear: both music and language stem from ancient communications that originally maintained and reinforced pair-bonds between mothers and infants. Not only were these primal duets crucial for the survival of prehistoric mothers and babies, they also enabled the survival of their descendants—among whom, happily, we count ourselves.

Additionally, a phenomenon called the Mozart Effect suggests that listening to music may increase intelligence. Although the

Mozart Effect has been controversial, there is evidence that formal music training (especially if begun early in life) improves long-term performance on spatial-temporal tasks and verbal memory.[68] The connection between music and memory seems almost obvious. For example, many children learn their alphabets from songs. Exposure to music is also good for babies' health. My colleague at Florida State University, Jayne Standley, has pioneered music therapy to help sick infants become healthy. She discovered that premature infants develop sucking abilities and gain more weight if they hear a female vocalist sing lullabies.[69] Altogether, these findings provide a clear message for parents: music is not a useless spandrel. Rather, it contributed to making us the intelligent, vocal species we have become, and—especially when expressed as lullabies and playsongs—continues to benefit our children today.

CHAPTER 8

ANCIENT GESTURES, MODERN ART

PAT-A-CAKE, PAT-A-CAKE

Pat-a-cake, pat-a-cake,

Baker's man!

So I do, master,

As fast as I can.

(Alternate clapping baby's hands and yours)

Pat it and prick it,

(Pretend to pat, then prick cake)

And mark it with B,

(Make a "B" in the air)

Put it in the oven

For baby and me.

(Alternate clapping baby's hands and yours)

—Mother Goose

Bᴏᴅʏ ʟᴀɴɢᴜᴀɢᴇ ɪꜱ ᴜꜱᴇᴅ universally but may take strikingly different forms in various human societies. Although many gestures, such as smiles, occur in all societies, some forms of body language are culture-specific. (I remember how surprised I was when I first saw someone in Puerto Rico pucker her mouth and move it sideways to surreptitiously point to something.) As we have seen, a person's body language or tone of voice can be even more important than the literal meaning of their words. Recall, also, how uncomfortable it is to interact with someone whose tone of voice is flat or whose facial expression is blank. Animated human bodies are fundamentally important for conveying information—so much so, in fact, that some believe that language evolved not from primate calls or songs but from gestures.

At some levels, this idea seems persuasive.[1] Although most primates regularly communicate with combinations of vocalizations, facial expressions, postures, and movements, only humans and great apes can also gesture with their free arms, hands, and (in the case of apes) feet.[2] Depending on the social contexts and traditions of a particular group, chimpanzees might extend their hands in gestures of begging or appeasement, raise their arms to make sexual advances, gently touch other individuals to request physical contact (such as nursing or sex), or clasp the hands of other apes overhead as a request for grooming. Orangutans, gorillas, and chimpanzees have more upright postures than most monkeys, and they stand up periodically, giving them a greater capacity for gesturing with their arms and hands.[3] This is partially why they can learn sign languages, to a limited degree.[4]

Because apes naturally have rich repertoires of gestures, it is generally assumed that our early ancestors did, too. As bipeds, our predecessors would certainly have had even freer forelimbs than apes, making gesturing easier. We can speculate more about the role of gestures in language evolution by studying the body language of young children because, according to the Haeckel Lite concept I discussed earlier, to some extent, the development of an individual mirrors the development of its species. As any parent knows, babies use gestures such as waving "bye-bye" before they produce their first words. Typically, children produce their first gestures between the ages of nine and twelve months, often by pointing to things, while the words for particular objects appear, on average, around three months after their respective gestures.[5] This suggests that words for objects or actions may also have emerged in our ancestors after they had developed gestural references for them.

Once they have begun to speak single words (often between the ages of ten and fourteen months), infants continue to produce gestures along with words—for instance, pointing to a cup while saying *cup*.[6] Gesture-word combinations that refer to one object precede gesture-word combinations that refer to two different concepts, such as pointing to a cup and saying *juice*. These latter combinations, in turn, occur before two-word combinations appear (between seventeen and twenty-three months).[7] Ultimately, changes in gesture not only predate but also predict changes in language. In other words, early gestures pave the way for language development in children.[8]

WHAT'S IN A GESTURE?

Unlike most words, gestures can be so transparent that one can understand (or invent) them without ever having been taught their meaning. A colleague based in Puerto Rico once told me about an English-speaking friend who wanted to order eggs for breakfast when he was visiting a restaurant there but didn't know the Spanish word for them (*huevos*). He solved the dilemma by pointing to his mouth, then standing up and squawking like a chicken while flapping his folded arms to simulate wings. He then made a plopping sound as he extended his hand down to catch an imaginary egg, which he presented to the waitress. The waitress understood perfectly. His pantomime was so transparent because he used gestures (flapping, cupped hand) and sounds (squawks) that physically resembled what they were supposed to represent. Such gestures that resemble their real-life counterparts are said to be iconic. Indeed, iconicity has been important in the evolution not only of gesture, but also of consciousness and both spoken and signed languages.[9] As the following paragraphs detail, it was also pivotal to the emergence of the visual arts and writing. That few nonhuman animals use iconicity shows that its emergence may have provided a cognitive breakthrough in our ancestors' face-to-face communication.[10]

Because of their freer forelimbs, hand-arm gestures would have been particularly important for our ancestors' face-to-face communication. Some of the earliest hand–arm gestures were probably iconic and contained elements of language, similar to contemporary hand–arm gestures that contain "embryo sentences."[11] For example, an iconic gesture meaning "catch" or "seize" consists of closing a hand around the extended forefinger of the other hand.[12] Not only is this gesture clear in its meaning, but it also contains a sentence: a subject (the moving hand), a verb (seizes), and a direct object (the extended forefinger).[13] Similar transparent gestures would have provided early hominins with a visual key, eventually leading to the development of syntax. And as we know, syntax was crucial for the emergence of language.[14]

We can make similar observations about gesturing in modern children. We all understand what a young child wants when he stands in front of an adult with his arms extended beseechingly overhead. In this case, the gesture together with the immediate environment (the nearby adult) contain the embryonic sentence "(You) pick me up!" Once an infant starts speaking, the word "up" (or "uppie") is frequently added to this gesture, resulting in a gesture–plus–word combination.

But not all gestures are iconic or even particularly transparent. Adam Kendon, a renowned expert on body language, has proposed that one way to sort gestures is along a continuum of precision, starting with those vague gestures that accompany speech (such as waving hands) to specific, conventional gestures used in formal sign languages.[15] David Armstrong, William Stokoe, and Sherman Wilcox, on the other hand, see a similar continuum but divide human gestures into four categories.[16] There is a basic primate level containing gestures that are very old in evolutionary terms, such as those associated with intimidation (looking bigger by puffing out the chest, holding the arms at a widened angle) or submission (drawing the arms in to appear smaller, cowering). The second category contains gestures that are understood everywhere by any person who is familiar with the objects or concepts to which they refer, but unlike the first category, they are unintelligible to nonhumans. (The classic hand gesture for a small gun is a good example of this kind of gesture.) The third level includes gestures that are not universal but are understood by all people living in certain societies, such as the pointing with puckered lips that I witnessed in Puerto Rico. Finally, the fourth level contains conventional gestures that are used in sign languages for the deaf but are not understood by people who are unfamiliar with the language.[17] Although Armstrong, Stokoe, and Wilcox's way of categorizing human gesture delineates more categories of gesture than Kendon's model, the two views are compatible and underscore the importance of gesture for language to emerge.

Sign languages for the deaf are of intense interest to scientists studying the link between gesture and the emergence of language, because sign languages are full-fledged natural languages with many of the same linguistic properties and constraints as spoken languages.[18] Although they are fully linguistic and therefore too sophisticated to be used as models for the first languages, sign languages nevertheless provide fascinating insights that are relevant for discussing the basic nature of language.[19] For example, the pioneering work of Susan Goldin-Meadow and her colleagues on deaf infants from America and China has shown that deaf children who have not been exposed to sign language *spontaneously* develop gestural communications that have formal properties of spoken languages.[20] This finding hints at an innate capacity for language, which other research pertaining to gestures and sign language supports.

For example, a remarkable new sign language has been created over the past thirty years by deaf children living in Nicaragua.[21] Before an elementary school for special education opened in Managua in 1977, sign language did not exist in Nicaragua. Instead, deaf children stayed at home, where they developed individual gestures ("home signs") to communicate with their families. Once the school opened, however, adolescent deaf children started to form a simple gestural system by combining elements from their various home signs.[22] Eventually, the Nicaraguan gestures expanded into an early sign language (just as pidgins can become established languages, or creoles). This early sign language, which eventually became known as Nicaraguan Sign Language (NSL), was then passed down to younger children who entered the community, and as second and third waves of youngsters learned the language, they began to transform it from a largely gestural system to a more complex sign language that had "quickly acquired the discrete, combinatorial nature that is a hallmark of language."[23] In addition, "The children who were arriving in the mid-1980s then became NSL's second wave of creative learners, picking up where the first cohort left off and making the changes that were never fully acquired by now-adolescent first-cohort signers. The difference today between first- and second-cohort signers therefore indicates what children could do that adolescents and adults could not. It appears that the processes of dissection, reanalysis, and recombination are among those that become less available beyond adolescence."[24]

Learning a foreign language with a perfect accent is much easier to do when you're very young. The Nicaraguan story suggests that this may also be true for inventing languages. In Chapter 6 we discussed that juveniles were largely responsible for disseminating sweet-potato washing and other inventions that were eventually handed down to future generations of Japanese macaques and that this remarkable finding is consistent with Jinyun Ke's computer simulations of how vocabulary may have emerged during prehistory. With these examples in mind, we can see how prehistoric children may well have played a crucial role during the development of protolanguage by inventing and disseminating early gestures and words.

So what can gesture—and sign language in particular—tell us about the importance of motherese? Primatologist Nobuo Masataka discovered that certain properties of motherese occur in American and Japanese sign languages when used with deaf infants. For example, similar to words in motherese, signs are delivered to deaf infants at a slower tempo and are exaggerated and repeated more often than signs used with deaf adults. Both hearing and deaf babies prefer to see infant-directed rather than adult-directed signs, which underscores Masataka's conclusion that "motherese is a prevalent form of language input to infants in speech or in sign."[25] Even more striking is that hearing infants who have had no exposure to Japanese Sign Language prefer viewing it in its motherese form, suggesting that infants' general preference for motherese (both signed and vocal) may be innate.

Another piece of gestural evidence demonstrates that the human capacity for language exists even in very young deaf infants. Laura Ann Petitto and her colleagues at Dartmouth recorded hand activities in two groups of babies when they were six months, ten months, and twelve months old.[26] One group consisted of three hearing babies who had not been exposed to sign language; the second included three hearing babies of profoundly deaf parents who had not received systematic exposure to speech but, instead, saw only signed language. The results were remarkable. Only the hearing babies of deaf parents "babbled" with their hands by moving them in rhythmic patterns similar to those of signing adults. Unlike the babies of the hearing parents, these infants also gestured mainly within a tightly restricted space that corresponds to mature signers' sign space.

These little babbling hands seem to be going through a kind of prelinguistic warm-up, similar to the vocal babblers I described in Chapter 5. Signed babble (like vocal babble) does not refer to anything specific. Instead, babies who babble by hand are practicing simple, discrete movements (gestures) as well as the rhythms for combining them that they have observed in adults. Once babblers have mastered the elements of gesture, they begin to form their first meaningful signs. After they get beyond the one-sign stage, babies begin to combine signs into communications that maintain the rhythms learned by babbling. Because signs are individual units, they can be combined and recombined into an infinite number of expressions—a universal feature of language.

Research on sign language is particularly important because it reveals that otherwise normal children who have been deprived of language are predisposed to create language spontaneously by observing communications around them. The process these deprived children go through is similar to what normal babies experience when they learn language. The deaf Nicaraguan children who had never been exposed to sign language extracted the most discrete embedded elements from the whole gestures they observed in others, and they eventually began combining these elements into sequences—much as younger infants naturally do when they babble and begin to produce and combine their first signs or words:

> Because NSL is such a young language, recently created by children, its changes reveal [two] learning mechanisms available during childhood. . . . The first is a dissecting segmental approach to bundles of information [that breaks] apart previously unanalyzed wholes. The second is a predisposition for linear sequencing; sequential combinations appear even when it is physically possible to combine elements simultaneously, and despite the availability of a simultaneous model. We propose that such learning processes leave an imprint on languages—observable in mature languages in their core, universal properties—including discrete elements (such as words and morphemes) combined into hierarchically organized constructions (such as phrases and sentences).[27]

In light of the NSL evidence, it is reasonable to suggest that as our ancestors' brains enlarged and evolved, prehistoric youngsters became capable of gleaning elementary bits of information from the expressions of others, including their mothers. These children likely expedited the development of protolanguage by honing extremely rapid subliminal abilities, not only for perceiving and analyzing the smallest fragments of movement or sound, but also for imitating and recombining them into meaningful sequences—and then spreading these sequences amongst their peers. While they developed NSL, the deaf Nicaraguan children also spontaneously invented clever syntax (for example, for expressing simultaneous actions).[28] It therefore seems likely that the syntax needed for arranging sounds and gestures emerged naturally as prehistoric children developed new ways to communicate.

IMITATION'S ROLE IN UNDERSTANDING OTHER MINDS

Speech requires fine movements of the vocal cords, throat, jaws, cheeks, tongue, mouth, and lips, and sign language incorporates precise movements of the fingers, hands, arms, shoulders, head, face, lips, and tongue. In turn, sensory perceptions complement motor behaviors, and recognizing objects or concepts precedes the ability to name, speak about, or use them.[29] However, as illustrated by the emergence of NSL, children must also be able not only to recognize and imitate behaviors, but also to comprehend what others *intend* by them to develop language.

The ability to imitate movements helps infants understand intention, according to Andrew Meltzoff, who studies newborns and young infants. As Meltzoff and his colleagues have shown, three-week-old babies can imitate certain tongue, lip, mouth, and finger movements of adults. This ability is inborn, says Meltzoff, and he may well be right, because studies of nonhuman primates suggest that infants' ability to imitate facial gestures may be very old in an evolutionary sense. For example, recent research shows that newborn monkeys and apes can imitate facial expressions, though it was once widely believed that they could not. Chimpanzees less than seven days old can imitate humans who stick out their tongues and open their mouths, although they lose the ability by the time they are two months old.[30] Three-day-old rhesus monkeys (macaques) imitate tongue protrusion and smacking of human lips but stop by the time they are two weeks old.[31] Pier Ferrari of the University of Parma in Italy and colleagues note that lip smacking is crucial for positive interactions among rhesus monkeys and suggest that newborns' imitation of this gesture reinforces mother-infant bonds and helps prepare infants for adult social life. They also have seen lip-smacking exchanges between rhesus mothers and their infants.[32]

Meltzoff argues that human infants' imitation of adult behaviors and their comprehension of what adults intend by those behaviors ultimately lead to a more generalized ability to infer others' mental states. This is significant because the ability to read the intentions and feelings of others (known formally as theory of mind) is thought to have been momentous during the evolution of human cognition. As Meltzoff notes, "In ontogeny, infant imitation is the seed and the adult theory of mind is the fruit."[33] But how do little babies ever learn to read others' intentions? To answer this question, Meltzoff looked at very young primates, including human babies, who have never seen their own faces because "there are no mirrors in the womb."[34] Although they cannot observe their own facial movements, they are still able to *feel* them. So when a newborn imitates an adult act, such as tongue protrusion, he associates the visual observation of the adult with the subjective feeling of performing the act, and this association is stored in memory.

Meltzoff believes human babies are born able to recognize the equivalence between acts they see and do, and that it allows them to map the relationship between their motor and mental experiences. The same process allows infants to infer the experiences of others. He explains, "when infants see others acting 'like me,' they project that others have the same mental experience that is mapped to those behavioural states in the self."[35] An important implication of this research is that infants' perception and production of motor behaviors are intertwined. Meltzoff speculates that the same kind of "*that*-looks-like-*this*-feels" process helps infants develop a theory of mind that incorporates visual, tactile, and motor senses.[36] As we have seen, motherese incorporates exaggerated facial expressions, and these expressions are probably easier for babies to imitate and feel. If so, in addition to helping infants acquire language, motherese contributes to their emerging abilities to infer mental states in others.

Before language could emerge, our ancestors had to replace many of their apelike communications with gestures and symbols that could be imitated, recombined, and used to generate a variety of meanings. Merlin Donald of Case Western Reserve University thinks hominins could not have done this until they first evolved an ability to refine or program their own motor skills.[37] So to develop variations of any action, including speech, it is necessary to rehearse the action, observe and remember its consequences, and then repeat and perfect the act in light of earlier results.[38] This mechanism, which Donald calls a rehearsal loop, resembles Meltzoff's description of motor imitation in young infants. Donald also notes that a precondition for the development of symbolic thinking is "a community of brains in interaction" that can produce new patterns of collective representations. Indeed, this "community of brains" existed among the deaf children in Nicaragua.[39]

Donald and Meltzoff persuasively argue that advances in our ancestors' intentional communications were not confined to any one modality but instead included numerous behaviors. Donald's rehearsal loops would have been important for the emergence of a variety of cultural behaviors, including song, dance, ritual, and, of course, language:

> Surely the most critical element is a capacity for deliberately reviewing self-actions, so as to experiment with them. Systematic and repetitive experimentation with action is evident fairly early in human development, especially in infant babbling, including manual sign-babbling. It would be no exaggeration to say that this capacity is uniquely human and forms the background for the whole of human culture, including language.[40]

GESTURES, VOICES, AND THE ORIGINS OF LANGUAGE

At this point, it is useful to review what we've learned so far: human communication is not just vocal, but also has a large component based on gestures. Be it a raised eyebrow, flicker across the eyes, shrug of the shoulders, fluttering of the hands, or tapping of the foot—all of these expressions are potentially rich sources of information about gesturers' mental states. As is the case with apes, the meanings of these gestures are usually general, rather than symbolic or linguistic. Human communication has evolved far beyond that of apes, however, because of our unique ability to invent and share bits of visual gesture or sound, string those bits together into an endless variety of meaningful sequences, and use them to communicate symbolically with others. The complement of this extraordinary skill is the ability to *understand* symbolic communications. As these abilities evolved, our ancestors also developed an ability to infer the mental states of others (theory of mind).

As I noted earlier, one school of thought believes language evolved from gestures rather than from primate calls or songs. Carried to its logical extreme, this hypothesis would mean the first true language was a sign language that speech somehow eventually supplanted. That seems unlikely, however, especially considering the multifaceted nature of social communication today. Language is incredibly complex—encompassing an enormous variety of gestures and vocalizations—and there is no reason to think that one of these modalities had priority over the other as language evolved in our ancestors.

Sherman Wilcox, a leading expert on linguistics and sign languages, also suggests that the debate about whether language evolved from gestures or vocalizations is based on an unnecessary dichotomy. Because there is abundant evidence that contemporary speech and gesture are linked, he observes that the two should not be separated in evolutionary accounts of language origins. We use finely controlled movements to broadcast information, whether we are gesticulating, signing, or speaking. In this sense, then, *all* languages are basically gestural in nature.[41] So "the remarkable human ability to acquire and use language regardless of modality does not depend on an abstract system of disembodied rules; rather, human language is the highly specialized, evolutionary manifestation of a multimodal gestural complex."[42]

In accordance with this theory, then, motherese is much more complicated than just "baby *talk*" or "musical *speech*." Not only do mothers speak to their babies in distinctive ways, but their baby talk is also accompanied by special gestures. For example, American and Italian mothers use a form of motherese in the gestures themselves, which are simpler and less abstract, and function to highlight and draw infants' attention to particular objects.[43] When teaching the names of objects to their babies, European, American, and Hispanic mothers speak, move the objects, and touch them to their infants' bodies.[44] Such special gestures reinforce the meaning of the mothers' words. Baby talk is also closely associated with mothers' facial expressions, which provide babies with additional information about verbal messages.[45]

In turn, babies' gestures and facial expressions provide moms with clues about their attentiveness and emotional state. Before infants produce their first words, their intentional gestures, such as pointing, may be accompanied by nonverbal vocalizations, such as cries or grunts.[46] As they mature, infants increasingly use combined gestures and vocalizations to achieve specific goals (rather than to express emotions), underscoring the importance of both gestures and vocalizations for the development of language.[47] By the time children are fifteen months of age, for example, the vast majority of their intentional pointing and reaching requests are accompanied by vocalizations.[48] Vocalization—in combination with gesture—primes babies' emerging verbal (linguistic) abilities. Ultimately, motherese, like the imitative underpinnings of signed and vocal languages, makes use of *multiple* modalities and, I believe, evolved incrementally from our ancestors' primarily emotional gestures and vocalizations to its present-day multifaceted form.[49] Today, motherese not only helps infants learn their first words, but also stimulates them to refine the imitative abilities needed to acquire words and theory of mind.

Human toddlers, unlike chimpanzees, eventually begin using gestures to communicate about ideas or objects that are not present, and their vocalizations become crucial for conveying the intentions behind their gestures.[50] Our proclivity for vocalizations about external or abstract ideas (instead of primarily self-centered ones) may have started when early hominin mothers began putting their babies down while foraging nearby. As mothers' and infants' voices began substituting for their clinging and cradling arms, their vocal communications started to take place over short distances. So for the first time in prehistory, young infants were routinely separated from their mothers, causing each to become part of the other's external world. Such separations would have been conducive to the emergence of intentional gestures about external events and ideas.

Refrigerator Art

Just as contemporary music seems to have evolved from emotional calls of our early ancestors, evidence points to the idea that drawings and paintings emerged from very ancient gestural systems.

Because they rely on applications (gestures) using paint or other media to capture and communicate static images, the visual arts may be thought of as embodying frozen gestures. Indeed, the flowering of children's artistic skills requires the same basic elements of multimodal processing that contribute to their verbal, musical, and gestural communications. Next we will explore how their artistic development might also parallel the emergence of artistic skills during human evolution.

To begin with, children ages eighteen months to two years practice the motor skills they will eventually use to draw or paint in a process similar to prelinguistic vocal and gestural babbling, but with scribbles. As with vocal and manual babbles, babies do not intend their initial scribbles to represent anything. Instead, scribbling youngsters are simply perfecting their motor coordination by moving markers in regular rhythmic motions.[51] As detailed by psychologists Glyn Thomas and Angèle Silk, rather than being aimless scrawls, these scribbles and patterns demonstrate a degree of visual balance and, over time, increasing hand–eye coordination.[52]

Young chimpanzees also scribble.[53] Both children and chimps become deeply absorbed and appear to take satisfaction in this activity. The main difference between the apes and children is that the former remain at the scribbling stage, while the children move on to produce more realistic drawings.

As children mature, so do their scribbles. Youngsters begin to interpret their completed scribbles as pictures, somewhat fancifully, by the time they are about two and a half years old.[54] Gradually, shapes such as separate lines, dots, and wobbly circles begin to appear among the scribbles. Just as vocally or manually babbling infants begin to combine the motor behavior they have mastered into their first words or signs, scribbling children start to incorporate these emerging shapes into their first pictures that appear meaningful to adults. This usually occurs by the time they are three to four years old. At first, children often fail to coordinate the parts of their images in a manner that makes sense to many adults. For example, body parts are often not attached in drawings of people. (To see examples, check the refrigerator doors of acquaintances with young children.) Soon, however, children learn to produce more coordinated images, just as they learn to build words or signs into appropriate phrases.

Indeed, children's first representative drawings frequently are of people, often composed, rather charmingly, of a circle for the head (with or without marks for the eyes, nose, or mouth) and two dangling lines for legs.[55] Arms, when included at this stage, commonly protrude from the sides of the head. In addition, a number of shapes appear in children's drawings across many cultures. These include circles, crosses, circles containing crosses, and rectangles. Despite the fact that human figures seem to be the most popular topic for young artists everywhere, houses and animals also have universal appeal.[56]

Basic shapes may gradually be combined and recombined to produce an endless variety of increasingly realistic pictures, a strikingly similar outcome to what happens after infants have mastered a certain number of words or signs. According to Thomas and Silk, European and North American children between the ages of five and eight often make X-ray drawings (depicting a baby in the mother's womb, for example). From

FIGURE 8.1. Josie Parkinson's portrait of her mother, which she drew when she was about three and a half years old.

FIGURE 8.2. A portrait by three-year-old Sisa Uzendoski. "This is my mom. She is a queen."

about eight years on, however, youngsters tend to produce more realistic images that incorporate depth, appropriate proportions, and a particular viewpoint. At around this age, many children begin to adopt conventional styles of drawing and often become dissatisfied with their art, which they abandon in favor of other forms of expression.[57]

Combining simple shapes and lines to produce pictures is a difficult task that entails planning, positioning, and sequencing.[58] The development of artistic skills in children requires sequential processing abilities and a grasp of how to combine visual elements into intentional images, similar to the requirements for language acquisition. Another parallel with language development is that very young scribblers understand some aspects of drawing before they are able to physically draw.[59] Kyoko Yamagata of Kanazawa University analyzed the processes underlying the emergence of representational drawing in eighty-seven children ranging from one to three years old. She found that when given a simple line drawing of a human or animal face and asked to color it in, children as young as eighteen months were able, to some degree, to color the individual components for eyes, nose, and mouth, although they could not have produced such representations without the template. Yamagata concluded that "the emergence of representational drawing during the scribbling stage is based on a process consisting of the extraction and drawing of component parts and the acquisition of a drawing method for organizing them."[60] Similar to speech comprehension, infants' grasp of the symbolic potential of drawing seems to emerge before their motor skills develop.

DOODLING IN THE DUST

When and how artistic abilities first emerged in our ancestors has interested scholars for many years. Some archaeologists believe art burst forth around forty thousand years ago (during the most recent or upper part of the Old Stone Age, or the Upper Paleolithic) in a "big bang" or "creative explosion" as a result of a relatively sudden genetic and/or neurological change in *Homo sapiens*. They also tend to agree with some linguists who suggest that language did not emerge from earlier primate call systems but, instead, evolved relatively recently and suddenly. For their part, these linguists frequently defend the hypothesis that language emerged recently by citing the recent Upper Paleolithic cave art from Europe. Gradualists like myself, on the other hand (and as I discuss in this chapter), think artistic abilities evolved slowly along with language over millions of years.

Artist and educator Susan Rich Sheridan, for example, believes artistic abilities emerged long ago within a multimodal framework that was strongly rooted in interactions between mothers and their offspring.[61] She envisions that solitary toddlers whose mothers had placed them on the ground engaged in a kind of prehistoric scribbling by "doodling in the dust." She also suggests that modern children produce marks similar to those early hominins made when they scribbled and drew. Using the natural unfolding of children's drawings as a model, she proposes that "early hominins scribbled first, drew schematically second, and then developed observational/representational drawing, thereafter inventing numbers, letters, algebra, calculus and musical nota-tion."[62] Although there is no way to find out whether hominin babies actually doodled in the dust (captivating thought though it is), we can study the archaeological record to get a rough idea of the sequence in which artistic skills may have emerged.[63]

Needless to say, our earliest ancestors did not paint or draw pictures on materials that lasted long enough to be posted on to-day's refrigerator doors. Our clues must come from modified rocks or fossilized artifacts, and there is not much of an archaeo-logical record of these until approximately two and a half million years ago. However, hominins certainly did not lack an aesthetic sense before that time. By at least three million years ago, australo-pithecines had acquired the habit of collecting fossils, crystals, or interestingly shaped pebbles and carrying them back to where they lived (these artifacts are now known as manuports).[64] Our early ancestors also appear to have had an eye for reddish rocks. That color was also favored by hominins who lived more than a million years ago and collected lumps of red ochre (a natural pig-ment that can be used to mark rocks) at Olduvai Gorge in Tanza-nia. (Over time, ochre of various hues became an important medium for aesthetic expression in many parts of the world.)

The earliest known artifacts that may have been associated with an emerging aesthetic sense were stone tools. These tools initially looked like clunky rocks but became increasingly refined as the Paleolithic slowly unfolded. Despite the tools' functional nature (often related to food), some of them suggest that well over two million years ago, our ancestors already appreciated a certain degree of symmetry, balance, and smoothness. Early tools from Africa, for example, included spheroid stones that some archaeol-ogists believe were used as hammers. Prehistoric art expert James Harrod notes, however, that some of these spheroids are simply too large to have been used as hammerstones. He suggests that, at least in these cases, they may simply have been aesthetically pleasing to our ancestors. In any event, the makers of the earliest known stone tools definitely had the conceptual machinery, motor skills, and inclination to sculpt objects out of rocks (if only roughly), representing a huge mental leap from their apelike ancestors.

FIGURE 8.3. Cupule and meandering
line from Auditorium Cave. *Photo courtesy
of Robert G. Bednarik.*

The earliest hominin-made artifacts were durable enough to
remain intact for millions of years because they were made of
rock. Because the vast majority of hominins' early creative efforts
in other materials would have disintegrated long ago, it would be
a mistake to assume that the "absence of evidence is evidence of
absence" for the emergence of art. It is also unreasonable to believe that the oldest known art is the first that existed. Rather, the
earliest known dates for artifacts or art are merely the most recent
dates *before which* they must have first appeared. How long before
those dates the earliest representatives of specific types of artifacts
or art appeared is usually unknowable, and earliest known dates
are often revised by new discoveries.

For instance, some of the oldest known examples of artistic
rock carvings (petroglyphs) are from Auditorium Cave in Bhim-
betka, India, and are dated to around 300,000 years ago.[65] The
cup-shaped mark and meandering line shown above were delib-
erately carved, which must have taken thought, motor skill, and
time. We will never know what these petroglyphs *meant* to their

maker, but the two shapes are similar to those in the early drawings of many contemporary children.[66] Did our ancestors spontaneously begin to make these well-formed marks, or were they
preceded in time by earlier, messier forms of art that may have
been closer to the scribbles of developing children? Archaeological evidence from a Lower Paleolithic *Homo erectus* site in Bilzingsleben, Germany, that was probably 400,000 to 300,000 years
old suggests the latter.[67] Bilzingsleben is exceptionally well preserved and has yielded over 100,000 artifacts, including wooden
staffs, polished ivory points, and a series of engraved bones from
large mammals such as elephants.[68]

Many of the engravings on the Bilzingsleben bones are multiple lines that converge, diverge, or, in some cases, remain parallel.
A few bones also have engraved arcs, double arcs, and lines at right
angles. On the whole, these designs appear to be simple and geometric. The eminent scholar of prehistoric art, Robert Bednarik,
has compared the markings from Bilzingsleben with engraved
specimens from more recent Middle Paleolithic sites in Europe
and elsewhere. This part of the Old Stone Age dates from approximately 300,000 to 40,000 years ago, and Bednarik thinks he sees
a "graphic evolution" from the convergence of many lines in
older specimens to patterns in which groups of lines are intentionally joined.[69] He also notes that "all of the markings of the Lower
and Middle Palaeolithic resemble modern doodling, which is
spontaneous and subconscious. Contemporary doodling, the scientific value of which remains almost entirely ignored, could well
have its neuropsychological roots in our early cognitive history."[70]

Bednarik has studied the earliest appearances of intentional
markings all over the world and was surprised that the same geometric designs that include simple lines, circles, dots, squiggles,
snails, crisscrosses, and so on appear universally. Humans are especially sensitive to these so-called *phosphene motifs* because they
cause particular groups of neurons to fire in the brain that are used
to process other more complex and dynamic visual stimuli.[71] You
may have experienced phosphene images when pressing on your
closed eyelids, when "seeing stars" after getting up too quickly, or
after a bump on the head. (I am periodically treated to an impres-

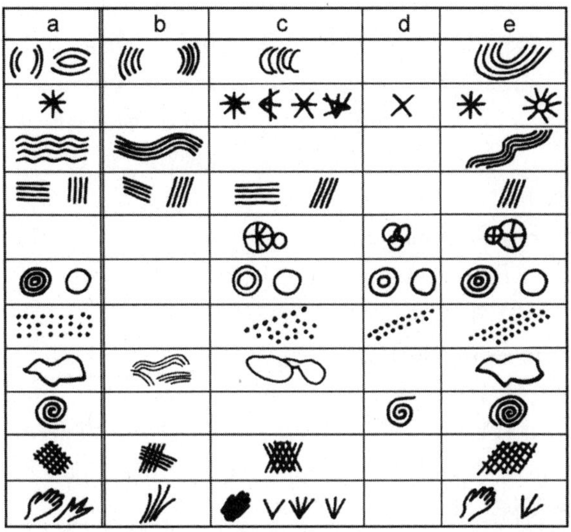

FIGURE 8.4. A comparison of (a) phosphene motifs with (b–e) motifs found in specific early graphic paleoart traditions. *Image courtesy of Robert G. Bednarik.*

sive and colorful array of them during optical migraines.) After discovering that phosphene designs appear repeatedly in both the earliest known prehistoric art and in the earliest drawings of children,[72] Bednarik proposed that

> the production of art-like markings commenced with such motifs, both in the modern individual and in the species as a whole. All drawings produced by infants up to the appearance of iconicity in their work (at about four years of age) have been noted to consist of a limited repertoire of phosphene motifs (Kellogg et al. 1965). I found that precisely the same applies to all motifs prior to the appearance of "pre-historic" figurative depiction. . . . All discoveries of the last twenty-six years have squarely confirmed my phosphene hypothesis, and no competing theory has stood the test of time.[73]

The phosphene table reproduced above was initially published by Bednarik in 1984.[74] Since then, many new discoveries have occurred, and some of the blank spaces have been filled. Although this table is twenty-five years old, I am including it here because I think Bednarik got it right. It is also worth noting that Bednarik's ideas about the emergence of phosphene motifs during prehistory are consistent with Sheridan's notion that earlier-living hominin babies scribbled in the dirt. Of course, phosphene images did not disappear from the archaeological record after representational drawings emerged. Instead, the hominin portfolio enlarged over time to include new types of art. Our ancestors continued to engrave or paint on rocks and make figurines, and later began to fashion jewelry and increasingly beautiful tools. Their favorite artistic subjects, remarkably, were the same as those of contemporary children around the world: human beings (frequently female) and animals.[75]

The candidate for the world's earliest-known iconic image that may represent a human figure is from Oldisleben, Germany. The image is engraved on bone, and the best estimate for its date is from 130,000 to 100,000 years ago.[76]

By the time of the Upper Paleolithic in Europe (forty thousand to ten thousand years ago), numerous caves contained images of geometric designs, human or mythical figures, stencils of hands, and, most dramatically, many animals.[77] Small portable sculptures also predominate in the Upper Paleolithic, including many charming "Venus figurines" of buxom women.[78] Beautifully engraved bone and antler tools had also emerged. The striking and sudden modernity of art is one reason some contend that an abrupt artistic explosion took place in Africa and Europe during this time. However, the earlier glimmers of artistry in the archaeological record outside of Africa are often overlooked.[79] Indeed, the accumulated evidence we discussed earlier strongly suggests that our ancestors gradually developed an increasingly complex facility for artistic expression over the long course of evolution—just as they did for language and music.

Small footprints and hand stencils from numerous caves that contain prehistoric art indicate that children were there, and some scientists are convinced that they had a greater part in creating cave art than previously believed.[80] Just as children probably

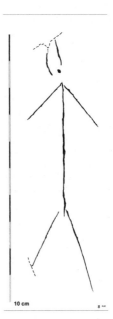

FIGURE 8.5. The engraving on the left from Oldisleben, Germany, may be the earliest known iconic image representing a human figure. The image on the right clarifies the engraving. *Images courtesy of Robert G. Bednarik.*

played a crucial part in the evolution of gesture and language, they may have influenced the development of human artistic sensibility. And the recent insights into children's gestures may have practical implications for modern parents. As I noted at the beginning of this book, my friend Betsy teaches her hearing babies sign language in addition to spoken words, which is a popular trend among parents today: "Teaching simple gestures, or signs, to babies before they can talk is a way to jump-start the language and communication process, and stimulate intellectual development. It can also confer a host of related benefits, including increased vocabulary, a deeper parent-child bond, enhanced self-esteem and decreased tantrums during the 'Terrible 2's,' proponents say."[81]

This notion of jump-starting communication skills in infants underscores an important point: although we have discussed children's acquisition of linguistic, musical, and artistic skills (and their appearances during evolution) in separate chapters, it is important to keep in mind that these abilities develop simultaneously in babies and that they depend on the maturation of similar neurological processes. Just as we concluded that it's important for mothers and fathers to speak and sing to their babies, parents may want to further boost their infants' communication skills by encouraging them to scribble, draw, finger paint, and tell stories about their art.[82] These artistic and gestural expressions are possible because of complex connections that develop as infants' brains mature. The complex interconnectedness of human brains, in turn, is a product of brain evolution, which is the topic of our last chapter.

CHAPTER 9

FINDING OUR TONGUES

IF WISHES WERE HORSES

If wishes were horses, beggars would ride.

If turnips were watches, I would wear one by my side.

And if "ifs" and "ands"

Were pots and pans,

There'd be no work for tinkers!

—*Mother Goose*

As I've made clear, I believe that human artistic skills evolved with, rather than after, language. Contemporary music can be considered "auditory cheesecake"[1] only in the sense that it appeared more recently than earlier musical forms and, therefore, came last. Similarly, we can view today's visual arts and modern languages as metaphorical "desserts" that were created from basic ingredients during our long evolution. So, too, was the human proclivity for inventing and improving everything from pots and pans to space stations and biotechnology.

With respect to the Mother Goose rhyme that opened this chapter, the invention of names was, of course, a crucial step that preceded the emergence of protolanguage, and the first words were probably nouns. If our ancestors never went beyond

161

the simple naming stage, they could not have subsequently invented the more subtle parts of speech ("... if 'ifs' and 'ands' were pots and pans"). But they did move beyond the naming stage due to the changes that occurred in the brain.

Although I have focused on mothers and infants as the primary targets of natural selection that eventually led to language, music, and art, I'm more inclusive in this chapter for an important reason: because of the way genes mix as they are transmitted to future generations, traits that are differentially selected for in one sex are extremely likely to affect the other. This is why people everywhere have a capacity for language and, to varying degrees, appreciate and participate in musical activities and the visual arts. Brain evolution, of course, was crucial for the emergence of these cognitive abilities in our ancestors.

Brain Evolution

The first time I saw a fresh brain being removed from a skull, I was shocked. The brains I had previously seen in neuroanatomy classes were solid, gray specimens that had been preserved in vats of fluid. I had dissected human brains in these classes and remembered that they had the consistency of well-done meatloaf. What a surprise, then, to see a *fresh* brain that wobbled like Jell-O—so much so that it would have slipped through the examiner's gloved fingers had it not been contained in a bag of membranes. Brains are about 78 percent water; hence their fluidity. Upon reflection, I realized that my astonishment stemmed in part from a previous assumption that an organ as complicated and important as the brain should also be substantial in a physical sense.

Human brains weigh, on average, a little less than three pounds, and they contain billions of nerve cells. Chimpanzees, on the other hand, have brains that weigh around three-quarters of a pound, which is very near the average for our earliest ancestors (the australopithecines), as estimated from their cranial capacities. Until recently, it was widely believed that at first, brain size increased gradually in australopithecines and then suddenly "took off" around two million years ago in their descendants (the earliest *Homo*), some of whom were our ancestors.[2] However, due to recent corrections in the cranial capacities and dates for certain fossils, a different picture is emerging.

As I noted in Chapter 4, fossils that are about 1.8 million years old from Dmanisi, Republic of Georgia, seem to have been transitional between those of australopithecines and early *Homo*. Although we are just starting to learn about body build in the Dmanisi hominins, their cranial capacities add to the growing doubt about a sudden takeoff in brain size around 2 million years ago.[3] Instead, graphs that include their cranial capacities along with corrected capacities and dates for other hominins suggest that the average brain size of our ancestors climbed more or less steadily for over 3 million years.[4] As a result, the average human brain is three times the size of an ape's, and it could evolve to be even larger in the future.[5]

This overall picture of brain size evolution fits with the gradual models I have described for the evolution of higher cognitive abilities, including language, music, and the arts. Thus, there is little evidence for a recent expansion in brain size. If anything, average brain size recently decreased, after the Neanderthals.[6] But the size of the brain does not explain everything related to cognition in our ancestors or in contemporary people. In fact, cranial capacities from normally functioning modern humans have been reported to vary enormously, from 790 to 2,350 cubic centimeters.[7] The relative sizes and connections of the brain's internal parts also evolved, and I believe that this so-called neurological reorganization was extremely important for human cognitive evolution.

From Grasping Hands to "Grasping" Brains

Neurological reorganization caused human brains to differ from those of nonhuman primates in a number of ways. For example, the degree to which the two sides of the brain control different activities is much greater in humans than in other primates. Although our two hands, like those of apes, are each controlled by the opposite side of the brain, most of us have strong preferences for using our right hands, unlike populations of nonhuman primates that generally favor neither. A region known as Broca's area on the left side of the human brain that is near the area that controls the right hand also predominates when it comes to speech. (This explains why right-handed people tend to gesticulate more with their right hands when they speak.) These facts are well known to scientists.[8] But other details are only now beginning to crystallize thanks to medical imaging techniques that make it possible to put a person in a machine (such as a functional magnetic resonance imaging [fMRI] scanner or a positron emission tomography [PET] scanner), ask him or her to think about something, and see what parts of the brain "light up."

Wonderful as these techniques are, they cannot be used to learn about how the internal organization of our ancestors' brains changed over time. We can get some idea, however, by comparing the brains of living apes and humans and examining endocasts from the braincases of fossil hominins. We know from apes and humans, for example, that the overall relative sizes of the major divisions (lobes) of the brain changed very little after hominins split from apes.[9] Thus, contrary to received wisdom, humans' frontal lobes are not especially large compared to the other lobes of the brain.[10] Rather, it is their neighbors, the temporal lobes, that appear to be a bit oversized—all the better for *Homo sapiens* to process speech, music, and other sounds; to identify people, animals, and objects; and to remember things. On the other hand, the relative size of the cerebellum, the great motor coordinator that sits underneath the back end of the brain, is somewhat reduced.

Studies comparing the brains of nonhuman primates and people also provide glimpses of how certain internal parts of the brain were rearranged as hominins evolved. For example, a deep structure that cannot be seen on the outside of brains, the insula, is somewhat enlarged in humans. This part of the brain is important for visceral sensations, taste, and certain aspects of speech.[11] In addition, the pattern of cells in a part of the visual area that is located at the back of the brain is advanced in humans in ways that may be related to decoding the rapid mouth and hand movements of speech and gesture.[12] Despite these observations, how-ever, neuroscience has revealed surprisingly little about the functional details of how our brains differ from those of other primates.

Todd Preuss of Yerkes National Primate Research Center, one of the foremost experts on primate brains, notes that it is not enough to compare brain tissue from dead monkeys and humans; what we need are comparisons of how brains actually function in live subjects, preferably chimpanzees and humans, which is now possible due to neuroimaging techniques. Preuss emphasizes that we need to break down behaviors into their smaller parts before we can unravel their evolution.[13] One example he offers involves the act of looking and reaching, which I find particularly compelling. After all, our ancestors' babies could no longer reach out and cling unsupported to their mothers. How, Preuss asks, could a seemingly simple act such as looking at an object and grasping it have been important for evolution?

He notes that many smaller parts go into this act: locating the desired object in space, discriminating its size, moving the head and eyes to see it more clearly, programming and carrying out the first movements of the hand toward it, adjusting the hand's movement as it zooms in, preshaping the hand to form a grasp, and adjusting the grip to accommodate the weight, compressibility, and texture of the grasped object. There is, of course, variation in the details: adjusting the shape of a grasping hand will be quite different in animals that clasp with their whole hands. As Preuss notes, "Once we understand the component processes that generate a particular behavior, or that comprise a particular psychological process, the scope of evolutionary changes one can conceive of (and thus explore empirically) is greatly enlarged."[14]

Evolution is a master tinkerer, as famously noted by cell geneticist Francois Jacob, because it builds new features by using old parts, and mounting evidence shows that language was likely built from networks in the brain that were originally important for reaching and grasping. Indeed, monkeys and apes rely largely on their hands to touch, feel, and explore individuals and objects, which is how they "grasp" their worlds. Although we depend heavily on language to grasp our worlds, vestiges of earlier evolutionary tinkering can be seen in the intertwined way in which our brains process language, on the one hand, and conceptualize graspable objects such as tools, on the other.[15]

The cartoon figure shown below (known as a homunculus) illustrates the general arrangement of the areas that receive sensations from the body and are located toward the back of the brain (on the right in this figure) and the motor areas in front that control its movements.[16] The homunculus in the left half of the brain represents the right side of the body (and vice versa) because brain connections are mostly crossed (except for parts of the face). For this reason, the homunculus here shows only right hands and feet. A pinched right thumb would be perceived in the thumb area toward the back of this illustration, while the thumb area in front of that would wiggle the thumb.

As illustrated, areas of the brain that allow people to "grasp" concepts of objects are close to the parts that facilitate grasping them with the hands. Thus, merely seeing a tool activates nerve

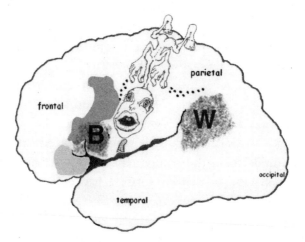

FIGURE 9.1. The left side of the brain, which is dominant for language, with the lobes labeled. Regions involved directly with language are Broca's area (B) and Wernicke's area (W). Merely seeing tools activates areas near the hand representations that would be involved in grasping them (dots on top); silently naming them lights up part of Broca's area (lower dots). The light gray area in front of B is activated during comprehension of spoken or signed sentences, whereas the dark gray region above B is important for grammar.[17]

cells near both the sensory and motor representations for hands (top dots in figure). When we think of how a tool is used, the activity increases in the dotted area that is nearest the motor area for the hand.[18] Further, if the tool is viewed and silently named, then part of Broca's speech area also lights up (the lower dots).[19] Clearly, the parts of the brain involved in seeing, thinking about, and naming tools are next to and, in some cases, partially overlap with the sensory and motor areas that facilitate actually grasping and manipulating them.[20] From the perspective of evolution, Preuss concludes that "the intimate relationship between language, object representation, and grasping is a particularly dramatic illustration of how evolution co-opts and modifies existing structures to serve new functions."[21] This offers a new twist to the metaphor of "grasping" meaning.

Recent imaging studies show that other nearby areas of the frontal lobe evolved to enable the understanding of subtle aspects of language.[22] Kuniyoshi Sakai of the University of Tokyo gives the following example: "John thinks that David praises his son" means something very different from "John thinks that his son praises David," although each sentence has the same exact words. A grasp of syntax, the part of grammar that arranges words into phrases and sentences, is necessary for this differentiation, and Sakai has clarified which parts of the left frontal lobe sort out these meanings. The light gray area in front of B in the homunculus in Figure 9.1 is selectively activated during comprehension of sentences (be they spoken or signed). But this region does not decode sentence meaning by itself. Rather, it communicates with the dark gray region above B in the figure, which Sakai regards as a grammar center that reflects the universal nature of grammatical processing. Although Sakai believes this grammar center has no counterparts in other animals, it's interesting that similar areas in monkey brains are important for orienting behaviors, because, in some sense, grammar (and syntax) are about orienting smaller units (such as words) within larger ones (such as sentences).[23]

If Preuss is right, then the saying "from hand to mouth" applies to much more than feeding. Indeed, our ancestors' babies lost the ability to reach out and cling to their mothers, and I have argued that this stimulated a long cascade of events that eventually resulted in the emergence of protolanguage. The parts of the brain involved in grasping and clinging would, naturally, have become modified as this happened, which fits nicely with the idea that hand representations were co-opted as language evolved. Instead of understanding an object merely by looking at it and manipulating it (or smelling it, or batting it with a paw) as other animals do, we have an added layer of comprehension that involves conceptualizing objects as bits of sound or signs (names) that may intentionally be transmitted to others. In addition, the brain anatomy needed for this added dimension may well have been cobbled together from networks in the brain that previously serviced looking and grasping.

THE BRAIN'S MIRRORS

Similar elements of brain anatomy can be seen today in monkeys, which provide the best living models available for unraveling human brain evolution. An area in the frontal lobe of the monkey brain becomes activated when they see other individuals reach out and grasp objects. In fact, certain nerve cells in this region fire both when a monkey sees an individual perform an action and when the observer does the same action—the ultimate in "monkey see, monkey do."[24] These mirror neurons, as they have been dubbed by their discoverers, Giacomo Rizzolatti and colleagues at the University of Parma in Italy, are quite remarkable because of this double capability. As Rizzolatti discovered, mirror neurons form a link between observers and actors by matching observed events to similar, internally generated actions in the observer that may (or may not) result in carrying out a similar action.[25] Mirror neurons are, thus, thought to play a role in the imitation, understanding, and learning of actions.

Interestingly, the part of the frontal lobe of monkeys that contains mirror neurons corresponds with part of Broca's speech area in humans, which is activated when we observe others moving their fingers and, even more so, when we actually imitate the behavior we see.[26] So we have mirror neurons, too.[27] Based on their work with mirror neurons, Rizzolatti and others believe that human language evolved from a basic mechanism that was originally related to the capacity to recognize actions in others and that manual gestures paved the way for speech to evolve. This idea meshes with the idea that evolution co-opted existing structures in our ancestors' brains to serve new functions.

But that's not all. Since the discovery of the first mirror neurons associated with reaching and grasping in the 1990s, neurons that mirror other kinds of stimuli have been discovered in monkeys and humans, and these may have been very important for the evolution of the cognitive abilities that set humans apart from other primates. For example, we have mirror neurons that are activated when we make or even just hear sounds that are associated with certain activities of the hand or mouth, such as ripping a piece of paper, popping open a can of soda, crunching a piece of candy with the teeth, kissing, and gurgling.[28] The activity that is provoked by sounds alone tends to be stronger in the left than the right side of the brain. Similar to the results for grasping compared to silently naming tools, the neurons that respond both to carrying out and hearing the sounds associated with hand actions are located nearer the hand areas in the brain, while those related to activities of the mouth are closer to its representation. In addition, when we merely listen to speech, parts of Broca's area that are activated when talking light up.[29]

There is other evidence that mirror neurons promoted mutual sensitivity between early hominin mothers and their infants and contributed to the evolution of their mutual communications. According to Stein Braten of the University of Oslo, when an infant chimpanzee grows old enough to shift from riding on its mother's belly to her back, it begins to see the world, literally, from her perspective. Thus, a back-riding infant "not only bodily moves with her movements but often adjusts its head to the mother's movement direction, thereby appearing to be gazing the same direction as the mother."[30] Unlike human babies, however, chimpanzee youngsters do not have prolonged eye contact with their mothers or much else in the way of face-to-face communication.

Early hominin infants who lost the ability to cling would have become extinct had they not evolved an ability to listen and learn by mirroring their mothers' actions. When infants shifted from their former piggyback perspectives to spending more time in their mothers' arms or on the ground, they became focused for the first time on face-to-face interactions.[31] This was likely a crucial step for the eventual emergence of language. Braten outlines additional steps that by now should be familiar to the reader: vocal imitation by newborns, turn-taking interactions with their mothers, selectively tuning in to sounds of their communities, and the emergence of babbling. As noted, Braten thinks these steps would not have come about without mirror neurons. The evolution of mirror neurons was also significant because it helped our ancestors develop a crucial ability for understanding the thoughts and motivations of other individuals.

WHERE'S TOM?

As I noted in the last chapter, the aptitude for knowing the emotional states, intentions, and motivations of others is called theory of mind, or TOM, and people are extraordinarily good at it. Professional dancers can rehearse routines in their minds and may also "feel" the performances that they see, which is an artistic, or even athletic, manifestation of TOM.[32] It should come as no surprise, then, that mirror neurons have recently been discovered near leg representations in the brain and that these respond when the opposite leg is gently stroked with a brush, as well as when people see this happening to others.[33] Christian Keysers from the University of Groningen calls the latter response "tactile empathy" and observes, among other interesting points, that it is why viewers of the James Bond film *Dr. No* feel their own bodies tingle disturbingly at the sight of a hairy tarantula crawling on 007's body. Our empathy is not just tactile, however. Viewers can tell from Bond's facial expression that he is frightened and, having read his mind, they are also frightened (and often experience racing hearts, shivers, and hairs standing on end).

Although other primates are not as skilled at using TOM as humans, they do not lack the ability altogether. Baboons, for example, have a certain amount of TOM. After years of attempting to understand baboon mental life by broadcasting recordings of their vocalizations from hidden speakers and observing the reactions of other baboons, it became clear to Dorothy Cheney and Robert Seyfarth at the University of Pennsylvania that the eighty or so baboons they studied knew one another's voices and understood social interactions just from hearing them.[34] Upon hearing recorded screams of a particular infant, for example, its mother looked toward the hidden speaker, while the other baboons looked at her. Some of the playbacks were edited to suggest highly unlikely social interactions, such as a threat grunt from a low-status baboon followed by a fearful scream from a dominant one. Compared with experiments with more realistic playbacks, these recordings caused surprise in the baboons. Cheney and Seyfarth concluded that baboons have TOM, but that it consists of vague intuitions about other animals' intentions rather than a humanlike ability for grasping specific goals, likes, dislikes, and motivations. This distinction is interesting because increased social awareness is thought to have played a crucial role during the evolution of human intelligence.

There is more to intelligence than just understanding social relationships, of course. For this reason, Esther Herrmann of the Max Planck Institute for Evolutionary Anthropology in Leipzig, Germany, and her colleagues recently investigated the relative amounts of knowledge about the physical world versus social intelligence in a large number of chimpanzees, orangutans, and human toddlers who were around two and a half years old.[35] Tests of physical-world knowledge included locating rewards after they were placed out of sight or rotated, discriminating quantities, and using a stick to obtain goodies that were out of reach. The social cognition tests included learning to solve a simple problem by watching a demonstration, understanding and producing gestures that indicate the location of hidden rewards, and two tasks related specifically to TOM—following an actor's gaze to an object and understanding an actor's intentions. The results were striking:

"Young children who had been walking and talking for about one year, but who were still several years away from literacy and formal schooling, performed at basically an equivalent level to chimpanzees on tasks of physical cognition but far outstripped both chimpanzees and orangutans on tasks of social cognition."[36]

Even more startling information about TOM has emerged in preverbal infants. Through a clever experiment that involved puppets, Kiley Hamlin and colleagues at Yale University demonstrated that babies as young as six months of age discriminate between characters that act nicely and badly toward others and, further, that they put the information to good use.[37] Hamlin began by showing infants a puppet show in which a circle with large button eyes tried in vain to climb a hill. Eventually the struggling climber was either helped by another good puppet or hindered by an evil one. Afterwards, infants were given a choice of reaching for either puppet, and almost all of them opted for good over evil. In fact, Hamlin was shocked by the strength of the responses, and thinks the ability to choose between nasty and nice individuals may be innate.[38] Perhaps the old adage about paying attention to children's and dogs' reactions to strangers has something to it.

These fascinating studies, taken together, suggest that the challenge of living in social groups may have been a driving force behind brain evolution in our ancestors—an idea that has existed for a long time and is tied to TOM. One version of this theory stresses the importance of competitive interactions involving social manipulation and deception, aptly referred to as Machiavellian intelligence.[39] A different view claims that brain size increased in ever-enlarging social groups to keep track of and manage increasingly tricky social relationships. In an interesting twist, this version of the theory suggests that language eventually emerged as a form of vocal grooming when group sizes became so large that there were too many acquaintances and too little time to groom them all in the old-fashioned, hands-on way.[40] This idea maintains that voices eventually began substituting for hands.[41]

Indeed, it seems that the evolutionary increase in hominin brain size was tied to emerging linguistic abilities. The following illustration of brain enlargement in growing humans during the

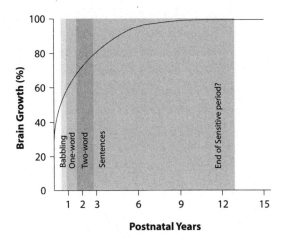

FIGURE 9.2. The increase in brain size during human development occurs during the years when children are most sensitive to acquiring language, which is consistent with the idea that brain enlargement may also have been related to the evolution of language during hominin evolution. (For each age, brain growth is shown as a percentage of the mature adult size.) *Image courtesy of Kuniyoshi Sakai.*

sensitive period from early infancy through puberty, when it is easiest to learn language, supports this theory.[42] The appearance of developmental milestones for speech such as babbling and formation of sentences occurs during the first years, when brain size is still small but its growth is fastest (reflected in the steepest part of the curve). Brain size levels off after the sensitive period for acquiring language. As I noted earlier, this contemporary pattern may reflect broad changes that occurred during the course of our ancestors' evolution.

Language would have sharpened TOM by permitting our ancestors to conceptualize such thoughts as "I believe that she knows where the fruit is, is waiting for me to leave before she gets it, and knows that's why I'm not leaving." Because mirror neurons become activated when individuals see other people having sensations and emotions as well as when they have identical experiences, it is likely that mirror neurons provide the underpinnings for TOM. Indeed, this idea is currently being verified by a small industry of neuroscientists who are using medical imaging technology to explore exactly where in the brain mirror neurons become activated when people consider the mental states of others. In an intriguing bit of converging evidence, people who score relatively high on tests that determine their degree of sensitivity to other people's perspectives (empathy) have the most active mirror neurons.[43]

COMING ATTRACTION

As if the information explosion medical imaging technology created were not enough, another revolution is bound to dramatically increase our understanding about brain evolution, and it is just in its beginning stages in a field called comparative genomics.[44] The long-awaited list of all the human genes (the human genome) was published in 2001.[45] The chimpanzee genome became available in 2005.[46] Before this, the assertion that humans and chimpanzees share about 99 percent of their genes had nearly achieved the status of fact. Now that the two genomes are available, however, scientists for the first time can compare how genes are internally organized, flipped, rearranged, deleted, or multiplied in the two groups. In reality, it's beginning to look like chimps and humans differ by about 4 percent rather than 1 percent in their overall genetic organization.[47] Contrary to common beliefs, we are actually not quite as close genetically to our chimpanzee cousins as we once believed. Nevertheless, they still remain, by far, our closest cousins.[48]

The study of genes that are associated with human diseases is also beginning to provide hints about the evolution of cognitive abilities in humans.[49] For example, the Forkhead box P2 (or *FOXP2*) gene came to attention because nearly half the members of a remarkable extended family were born with a developmental disorder that severely limited their speech. Imaging studies showed that, when speaking, family members with mutations in this gene had less activity than normal relatives in and near Broca's speech area on the left sides of their brains.[50] Instead, their brains lit up in more posterior regions on both sides. Although it has been suggested that the normal human form of the *FOXP2* gene evolved recently and provided a kind of "magic bullet" that caused a sudden emergence of language, this seems unlikely. Variants of the *FOXP2* gene appear in animals such as mice and birds, and the human gene is expressed in a variety of tissues including the developing heart, lungs, and gut.[51] *FOXP2*, therefore, probably has a more general function than simply being a language gene. As Michael Corbalis from the University of Auckland, New Zealand, observes, "The evolution from non-speaking ape to talkative human must have involved multiple genes and many different steps."[52] Although new, the melding of knowledge about the genetics of human diseases with information about brain function looks very promising for future studies of hominin brain evolution.

MOTHERS AND INFANTS—PAST AND FUTURE

While it is common knowledge that our children are our species' future, this book shows how they, and their mothers, were also crucial for our past. Natural selection sculpted the human journey, as it always does, by determining who lived and who died. Because our early ancestors behaved much like modern chimpanzees and had no concept of paternity, the onus of raising healthy babies was mostly on prehistoric mothers. When our ancestors began walking on two legs, the changes in their bodies made mothers and infants lightning rods for natural selection. This alone is justification for exploring the impact of mothers and infants on our extraordinary evolution. Another reason is that the theories that have traditionally addressed human evolution have usually focused on men to the exclusion of almost everyone else. Despite the fact that grandmothers and food-gathering mothers have begun to receive a share of the credit for our species' remarkable achievements, children have been ignored.

But no longer. Although at first blush it seems like such a simple thing, a diminished grasping ability in prehistoric babies who survived the evolutionary shift to bipedalism profoundly affected human evolution. Little ones who could no longer cling to their mothers began to protest vociferously when they were put down nearby so that moms could collect food or simply rest. For the first time in prehistory, mothers began responding voice-to-voice by inventing the first lullabies and soothing motherese as part-time replacements for cradling arms. Later, our ancestors' first words and, eventually, their first language were among the spinoffs of these innovations and, as the recent emergence of Nicaraguan Sign Language illustrates, these breakthroughs were probably invented and spread largely by children. And all because feet that had become tailored for walking and helpless little hands lost the ability to cling.

Of course, the hands of baby hominins never lost their *inclination* to grasp. In fact, it's still there in our newborns, as shown in the photograph of my day-old grandson on his father's arm. But the ability to cling unassisted was irretrievably lost. Although baby hominins that survived this natural experiment were slower to develop motor skills than their ancestors' babies had been, that doesn't mean they grew up to be clumsy. On the contrary, the hominin hand changed over time in a way that made a precision organ even more precise. In the process, thumbs became very flexible and more useful for picking berries, making better tools, and (eventually) holding color crayons.

Evolutionary changes in the anatomy and functions of hands and feet corresponded with changes in their representations in the brain. Just look at the size of the hands and thumbs on the homunculus in the figure we've seen. And isn't it beautiful that Mother Nature appears to have co-opted networks of the brain that were (and are) involved in reaching for and grasping objects when she gave us the linguistic facility for "grasping" concepts? This is especially poignant because it was the grasping hands and feet—or, rather, the lack thereof—that necessitated the development of motherese and, ultimately, the emergence of language itself. What goes around comes around!

As we have seen, art also developed in lockstep with the brains of our ancestors. So did music. Their basic ingredients were similar and consisted of discrete elements, such as musical notes or graphic marks, that could be recombined into larger meaningful units and shared with others, as well as rules (syntax) for doing so. These are the same elements that our ancestors used to invent language. Ultimately, if our ancestors' babies had not lost the ability to cling to their mothers, there would be no Mozart piano concertos and no Mona Lisa. No ballet, no Shakespeare, and no $E=MC^2$. Indeed, if it weren't for the first coos of motherese, spoken to soothe our earliest ancestors, there would be no humans as we know them today.

Clearly, our species owes a special debt not only to prehistoric mothers, but also to their children who were the impetus for the evolution of the cognitive features that set us apart from other animals. It is amazing to contemplate that language and the conscious thoughts that it permits evolved, ultimately, from prehistoric infants' simple need to be held and comforted by caregivers. Harlow's earlier experiments on monkeys and myriad more contemporary studies on primate infants, including humans, show that youngsters' need for physical contact with parents or other caregivers is profoundly important for primates. If the past is the best predictor of the future, human parents everywhere would do well to pay more attention to this need.

ACKNOWLEDGMENTS

I AM DEEPLY GRATEFUL to my agent, Deirdre Mullane, and to William Frucht for deciding that this book was a worthwhile project. I thank my editor at Basic Books (Perseus), Amanda Moon, and Alix Sleight, who provided thorough feedback for improving the initial manuscript. I also appreciate the hard work of Sandra Beris, project editor, Antoinette Smith, and Wesley Weissberg at Basic Books. In addition, the following people kindly provided helpful correspondence and discussions, images, and moral support:

Myrdene Anderson; Robert G. Bednarik; young Christopher and his parents; Colette Burbesque; Curt Busse; Ron Clarke; Frederick Coolidge; Inge Cordes; Alyssa Crittenden; Kylene Dolen; Ellen Dissanayake; Lance, Jocelyn, and Grant Gravlee; Harlow Primate Laboratory; Charles Hildebolt; Helen A. LeRoy; Frank Marlowe; National Geographic Society; Nobuo Masataka; Werner Mende; NASA Johnson Space Center; Chisako Oyakawa; William, Betsy, and Josie Parkinson; Michael, Adrienne, and Jacob Riddle; Kuniyoshi Sakai; Sarah Schofield; Susan Sheridan; Chris Sloan; Kirk Smith; Laurel Trainor; Wenda Trevathan; Michael, Edith, and Sisa Uzendoski; Kathleen Weinke; Nadine Woihsa; Joel and Lisa Yohalem; and Adrienne Zihlman.

I have been privileged to be a resident scholar at the School for Advanced Research (SAR) in Santa Fe, New Mexico, during the 2008–2009 academic year, and will be eternally grateful to James Brooks and the other powers-that-be at SAR for this wonderful experience.

NOTES

CHAPTER 1: SILENCE IS GOLDEN

1. Adolph H. Schultz, *The Life of Primates* (New York: Universe Books, 1969).

2. Jane Goodall, *Through a Window* (Boston: Houghton Mifflin, 1990).

3. Frans de Waal, *Good Natured* (Cambridge, MA: Harvard University Press, 1996), pp. 19–20.

4. Schultz, *The Life of Primates*.

5. Jane Goodall, *The Chimpanzees of Gombe: Patterns of Behavior* (Boston: Belknap Press of Harvard University Press, 1986).

6. Goodall, *Through a Window*, pp. 39–40.

7. Ibid., p. 161.

8. Chris Case, "All That's Glitter and Golden: The Story of Gombe's Famous Twins," Jane Goodall Institute News Center, 2005, www.janegoodall .org/news. (Enter "Glitter" in search box.)

9. This is not to say that chimpanzee and human babies are not similar in much of their development. Compared to those of monkeys, the periods during which young chimpanzees and humans are emotionally and physically dependent upon their mothers are extended. This gives youngsters more time to acquire the social skills and learning associated with the bigger brains, greater intelligence, and longer lives that evolved in great apes and, even more so, in humans. In addition to helplessness at birth, we share other remarkable similarities with chimpanzees in the timing of the first occurrences of certain developmental milestones, such as the disappearance of blind rooting for the nipple, distress at separation from the mother, social faces (e.g., smiling in humans or requesting to play by making a "play face" in chimpanzees), and fear of strangers (Frans X. Plooij, *The Behavioral Development of Free-living Chimpanzee Babies and Infants* [Norwood, NJ: Ablex, 1984]). Significantly, humans and chimpanzees also differ in the timing of other landmarks that are related not just to grasping, but also to broader aspects of posture and locomotion. Pushing off, sitting without support, creeping on all fours, standing without support, and walking occur much later in human babies.

10. Goodall, *Through a Window*, p. 170.

11. Anne Pusey, Jennifer Williams, and Jane Goodall, "The Influence of Dominance Rank on the Reproductive Success of Female Chimpanzees," *Science* 277 (1997): 828–831.

12. Sarah Blaffer Hrdy, *Mother Nature: Maternal Instincts and How They Shape the Human Species* (New York: Pantheon, 1999), p.161. See also Craig B. Stanford, "To Catch a Colobus," *Natural History* 104 (January 1995): 48–55.

13. Bonobos have just recently been reported to hunt and consume monkeys, which, previously, they had been thought not to do: Gottfried Homann and Barbara Fruth, "New Records on Prey Capture and Meat Eating by Bonobos at Lui Kotale, Salonga National Park, Democratic Republic of Congo," *Folia Promatology* (Basel) 79 (2008): 103–110; Martin Surbeck and Gottfried Hohmann, "Primate Hunting by Bonobos at Lui Kotale, Salonga National Park," *Current Biology 18* (2008): R906–907.

14. Goodall, *Through a Window,* p. 34.

15. Goodall, *The Chimpanzees of Gombe,* p. 582.

16. Takayoshi Kano, *The Last Ape* (Stanford: Stanford University Press, 1992), p. 165.

17. Goodall, *Through a Window,* p. 197.

18. Ibid.

19. Michael Tomasello and Luigia Camaioni, "A Comparison of the Gestural Communication of Apes and Human Infants," *Human Development* 40 (1997): 7–24.

20. Goodall, *The Chimpanzees of Gombe.*

21. Much of this information is from Shozo Kojima, *A Search for the Origins of Human Speech* (Kyoto, Japan: Kyoto University Press, 2003). This book is a must-read for researchers who are interested in mother-infant communication in apes.

22. Nadezhda Nikolaevna Ladygina-Kohts, *Infant Chimpanzee and Human Child: A Classic 1935 Study of Ape Emotions and Intelligence*, Frans B. de Waal, ed. (Oxford, UK: Oxford University Press, 2002).

23. Ibid., p. 205.

24. Ibid., p. 84.

25. Ibid., p. 118.

26. Ibid., p. 221.

27. Ibid., p. 199.

28. Ibid., p. 201.

29. To learn more about Nadia Kohts's remarkable career, see Frans B. de Waal, "Before Jane Goodall, There Was Nadia Kohts," *Chronicle of Higher Education* 49, no. 15 (2002): B11.

30. Plooij, *The Behavioral Development of Free-living Chimpanzee Babies and Infants.*

31. Nancy A. Nicolson, "A Comparison of Early Behavioral Developments in Wild and Captive Chimpanzees," in *Primate Bio-social Development,* Suzanne Chevalier-Skolnikoff and Frank E. Poirier, eds. (New York: Garland, 1977).

32. Christophe Boesch and Hedwige Boesch-Achermann, "Dim Forest, Bright Chimps," *Natural History* 100 (September 1991): 50–57.

33. Goodall, *The Chimpanzees of Gombe*, pp. 369–370.

34. Kano, *The Last Ape,* p. 132.

35. Plooij, *The Behavioral Development of Free-living Chimpanzee Babies and Infants*, p. 142.

36. Kojima, *A Search for the Origins of Human Speech*, p. 139.

37. Goodall, *Through a Window*, p. 163.

38. Goodall, *The Chimpanzees of Gombe,* p. 131.

39. Dario Maestripieri and Josep Call, "Mother-Infant Communications in Primates," *Advances in the Study of Behavior* 25 (1996): 613–642.

40. Kojima, *A Search for the Origins of Human Speech*.

CHAPTER 2: SAYING WHILE SOOTHING

1. Frank W. Marlowe, "Hunter-Gatherers and Human Variation," *Evolutionary Anthropology* 14 (2005): 54–67.

2. Adolph H. Schultz, *The Life of Primates* (New York: Universe Books, 1969), p. 152.

3. Specifically, the entire pelvis of bipedal hominins became compressed vertically, so it appears short and squat compared to that of a great ape. The flaring blades of the pelvis also curved around from the back to the side, creating a bowl-like shape. Meanwhile, the spine changed with upright walking, so the tailbone curves forward near the bottom of the pelvis, adding to the bowl effect. The result of these modifications is a human pelvis that constricts the birth canal in women.

4. Sarah Blaffer Hrdy, *Mother Nature: Maternal Instincts and How They Shape the Human Species* (New York: Pantheon Books, 1999), p. 471.

5. Stephen Jones, "Natural Selection in Humans," in *Cambridge Encyclopedia of Human Evolution,* S. Jones, Robert Martin, and David Pilbeam, eds. (Cambridge, UK: Cambridge University Press, 1992), p. 286.

6. Judy DeLoache and Alma Gottlieb, eds., *A World of Babies, Imagined Childcare Guides for Seven Societies* (Cambridge, UK: Cambridge University Press, 2000).

7. Hrdy, *Mother Nature,* p. 476.

8. Ibid., pp. 477–484.

9. According to a recent study of 1,254 newborns, "postnatal weight loss is a well known but little studied phenomenon. It represents mainly fluid loss but may also involve loss of fat stores during the establishment of milk feeding." Charlotte M. Wright and Kathryn N. Parkinson, "Postnatal Weight Loss in Term Infants: What Is 'Normal' and Do Growth Charts Allow For It?" *Archives of Disease in Childhood, Fetal and Neonatal Edition* 89 (2004): p. 255.

10. Women give birth to babies that are more than twice the weight of newborn chimps, but, then, women themselves are bigger than adult female chimpanzees. Because the newly delivered human baby's brain is twice the size of the newborn chimp's (about 350 versus 150 cubic centimeters), relative brain sizes (brain size divided by body size) are about the same at birth for the two species. Both chimpanzees and people may be born when their brain weights reach approximately 10 percent of their body weights. This fraction decreases, however, as individuals grow up—think of how odd grown-ups would look if their heads were as relatively large as those of infants! The similarity in relative brain size at birth is remarkable because relative brain size of mature humans is much larger than that of adult chimpanzees. Compared to chimpanzee newborns, human babies experience tremendous brain growth during their first few

years of life, which is why the brains of adult humans are three times the size expected for an ape of equivalent body size.

11. Hrdy, *Mother Nature,* p. 481.

12. For details, see Karen Rosenberg and Wenda Trevathan, "The Evolution of Human Birth," *Scientific American* 285 (November 2001): 72–77, including a fascinating description of a woman who was stranded in a tree during a flood giving birth there.

13. DeLoache and Gottlieb, eds., *A World of Babies.*

14. Le Huynh-Nhu, "Never Leave Your Little One Alone, Raising an Ifaluk Child," in DeLoache and Gottlieb, eds., *A World of Babies.*

15. Alma Gottlieb, "Luring Your Child into This Life: A Beng Path for Infant Care," in DeLoache and Gottlieb, eds., *A World of Babies.*

16. Carol Delaney, "Making Babies in a Turkish Village," in DeLoache and Gottlieb, eds., *A World of Babies.*

17. Marissa Diener, "Gift from the Gods, a Balinese Guide to Early Child Rearing," in DeLoache and Gottlieb, eds., *A World of Babies.*

18. Sophia L. Pierroutsakos, "Infants of the Dreaming, a Warlpiri Guide to Child Care," in DeLoache and Gottlieb, eds., *A World of Babies.*

19. Ibid., p. 160.

20. Michelle C. Johnson, "The View from the Wuro, a Guide to Child Rearing for Fulani Parents," in DeLoache and Gottlieb, eds., *A World of Babies.*

21. Ibid., p. 179.

22. Marlowe, "Hunter-Gatherers and Human Variation," p. 62.

23. Meredith F. Small, *Kids, How Biology and Culture Shape the Way We Raise Young Children* (New York: Anchor Books, 2001), p. 204.

24. Hrdy, *Mother Nature,* p. 519.

25. Sarah Blaffer Hrdy, "Fitness Tradeoffs in the History and Evolution of Delegated Mothering with Special Reference to Wet-Nursing, Abandonment, and Infanticide," in *Infanticide and Parental Care,* Stefano Parmigiani and Frederick S. vom Saal, eds. (London: Harwood Academic, 1994).

26. Joseph Soltis, "The Signal Functions of Early Infant Crying," *Behavioral and Brain Sciences* 27 (2004): 443–490.

27. DeLoache and Gottlieb, eds., *A World of Babies.*

28. Johnson, "The View from the Wuro"; Huynh-Nhu, "Never Leave Your Little One Alone."

29. Gottlieb, "Luring Your Child into This Life."

30. Diener, "Gift from the Gods."

31. Delaney, "Making Babies in a Turkish Village."

32. Gottlieb, "Luring Your Child into this Life," pp. 80–81.

33. John Bowlby, *Attachment and Loss, Vol. 1: Attachment,* 2nd ed. (New York: Basic Books, 1982).

34. Ann Frodi, "When Empathy Fails," in *Infant Crying: Theoretical and Research Perspectives,* B. M. Lester and C. F. Z. Boukydis, eds. (New York: Plenum, 1985).

35. Kim A. Bard, "What Is the Evolutionary Basis for Colic?" *Behavioral and Brain Sciences* 27 (2004): 459.

36. Soltis, "The Signal Functions of Early Infant Crying," p. 454.

37. Ibid., pp. 443–490.

38. Ibid., p. 450.

39. Ibid., p. 453.

40. Irenäus Eibl-Eibesfeldt, *Human Ethology* (New York: Aldine de Gruyter, 1989), pp. 193–194.

41. Meredith Small, *Our Babies, Ourselves* (New York: Anchor Books, 1998), p. 156.

42. Small, *Our Babies, Ourselves,* and Peter H. Wolff, "The Natural History of Crying and Other Vocalizations in Early Infancy," in *Determinants of Infant Behavior, Vol. 4,* B. M. Foss, ed. (London: Methuen, 1969).

43. K. Christensson, T. Cabrera, E. Christensson, K. Uvnas-Mohbery, and J. Winberg, "Separation Distress Call in the Human Neonate in the Absence of Maternal Body Contact," *Acta Paediatrica* 84 (1995): 468–473.

44. Silvia M. Bell and Mary Dinsmore Salter Ainsworth, "Infant Crying and Maternal Responsiveness," *Child Development* 43 (1972): 1171–1190.

45. Bard, "What Is the Evolutionary Basis for Colic?"

46. H. M. Halverson, "Studies of the Grasping Responses of Early Infancy: I," *Journal of Genetic Psychology* 51 (1937): 371–392.

47. DeLoache and Gottlieb, eds., *A World of Babies.*

48. Nicholas G. Blurton-Jones, "The Lives of Hunter-Gatherer Children: Effects of Parental Behavior and Parental Reproductive Strategy," in *Juvenile Primates,* M. E. Pereira and L. A. Fairbanks, eds. (New York: Oxford University Press, 1993).

49. Patricia Draper, "Social and Economic Constraints on Child Life Among the !Kung," in *Kalahari Hunter-Gatherers,* R. B. Lee and I. DeVore, eds. (Cambridge, MA: Harvard University Press, 1976).

50. For discussion, see Small, *Our Babies, Ourselves.*

51. Gottlieb, "Luring Your Child into This Life," p. 83.

52. Hrdy, *Mother Nature,* p. 503.

53. Gottlieb, "Luring Your Child into This Life," p. 82.

54. Diener, "Gift from the Gods," p. 109; Johnson, "The View from the Wuro," p. 185.

55. Pierroutsakos, "Infants of the Dreaming," p. 163.

56. Gottlieb, "Luring Your Child into This Life," p. 86.

57. Robert R. Provine, "Walkie-Talkie Evolution: Bipedalism and Vocal Production," *Behavioral and Brain Sciences* 27 (2004): 520–521. See also Robert R. Provine, *Laughter* (New York: Penguin Books, 2000),

58 Ellen Dissanayake, "Antecedents of the Temporal Arts in Early Mother-Infant Interaction," in *The Origins of Music,* N. L. Wallin, B. Merker, and S. Brown, eds. (Cambridge, MA: MIT Press, 2000), p. 391.

59. Sandra E. Trehub, Laurel J. Trainor, and Anna M. Unyk, "Music and Speech Processing in the First Year of Life," *Advances in Child Development* 24 (1993): 1–35. The authors note that soothing songs are sung less in North America and Europe, where caregivers typically withdraw from infants' rooms before they fall asleep.

60. Sandra E. Trehub, Anna M. Unyk, and Laurel J. Trainor, "Adults Identify Infant-Directed Music Across Cultures," *Infant Behavior and Development* 16 (1993): 193–211.

61. Eibl–Eibesfeldt, *Human Ethology*.

62. Modified slightly from L. J. Trainor, E. D. Clark, A. Huntley, and B. A. Adams, "The Acoustic Basis of Preferences for Infant-Directed Singing," *Infant Behavior and Development* 20 (1997): 383–396.

63. Trainor, Clark, Huntley, and Adams, "The Acoustic Basis of Preferences."

64. L. J. Trainor, C. M. Austin, and R. N. Desjardins, "Is Infant-Directed Speech Prosody a Result of the Vocal Expression of Emotion?" *Psychological Science* 3 (2000): 188–195.

CHAPTER 3: AND DOWN WILL COME BABY

1. Frans X. Plooij, *The Behavioral Development of Free-Living Chimpanzee Babies and Infants* (Norwood, NJ: Ablex, 1984).

2. "Hominin" means what "hominid" used to mean (i.e., our bipedal ancestors and ourselves, but not apes), although some people still prefer to use "hominid" for this. The more contemporary use of "hominid," however, now includes great apes. Because there is no obvious way of knowing which meaning of "hominid" is intended, and because the term "hominin" has only one meaning—not to mention that it is now the preferred term for our bipedal relatives and ourselves—it is used in this book.

3. Harry F. Harlow, "The Nature of Love," *American Psychologist* 13 (1958): 678.

4. John Watson, *Psychological Care of Infant and Child* (New York: Norton, 1928), pp. 81–82.

5. Meredith F. Small, *Our Babies, Ourselves* (Anchor Books, 1998); Peter H. Wolff, "The Natural History of Crying and Other Vocalizations in Early Infancy," in *Determinants of Infant Behavior,* B. M. Foss, ed. (London: Methuen, 1969).

6. H. M. Halverson, "Studies of the Grasping Responses of Early Infancy: I," *Journal of Genetic Psychology* 51 (1937): 371–392.

7. Webster's dictionary defines "experiment" as "any action or process undertaken to discover something not yet known or to demonstrate something known." This definition implies a conscious experimenter exists, of course, which the term "natural experiment" does not. Instead, the latter refers to a naturally occurring event or process from which we may gather further evidence about something or learn something new. For example, the incident in which Madam Bee was stricken with polio and then gave birth to Bee-hind was a natural experiment. Although Madam Bee's first two offspring thrived, Bee-hind died shortly after birth because she could not stay attached to her mother, which suggests that sustained contact between newborns and their mothers is vital not just for monkeys, but also for apes.

8. Amazingly, Darwin and Wallace formulated their theory without knowing about genes. Today, of course, we do know about them thanks to Gregor Mendel. Genes are the basic stuff of evolutionary change and, as we have seen, they are disseminated by successful reproduction.

9. Although myriad comparative genetic and molecular studies and evidence from the fossil record converge on this estimate for the timing of the divergence of the hominin and chimpanzee lines, murmurings are just emerging in the field that this date may in fact be too recent. Stay tuned!

10. Dennis M. Bramble and Daniel E. Lieberman, "Endurance Running and the Evolution of *Homo*," *Nature* 432 (2004): 345–352.

11. Pete E. Wheeler, "Stand Tall and Stay Cool," *New Scientist* 12 (May 12, 1988): 62–65.

12. Since apes are the best living models for early hominins, giving birth must have been uncomplicated before the irresistible force of increasing brain size met the immovable object of pelvic modifications for bipedalism. Apes have pelvic inlets that are spacious compared to the heads of their neonates, so their births are quick and easy. Ape newborns' bodies are tiny compared to adults, so they quickly catch up after they are born. This rapid growth partially explains why they soon develop motor skills that allow them to cling autonomously to their mothers. Their brains, however, do not increase nearly as much as those of human newborns.

13. P. Brown et al., "A New Small-Bodied Hominin from the Late Pleistocene of Flores, Indonesia," *Nature* 431 (2004): 1055–1087; M J. Morwood et al., "Archaeology and Age of a New Hominin from Flores in Eastern Indonesia," *Nature* 431 (2004): 1087–1091.

14. The nature of birth remains an open question for the much more recently living species, *Homo floresiensis*, since the brain size of LB1 (the museum number for Hobbit) was in the australopithecine range; Dean Falk et al., "The Brain of LB1, *Homo floresiensis*," *Science* 308 (2005): 242–245.

15. A third hominin, *Paranthropus* (robust australopithecines), also lived around 1.5 million years ago, but this group is rarely entertained as a possible direct ancestor of humans.

16. Alan Walker and Christopher B. Ruff, "The Reconstruction of the Pelvis," in *The Nariokotome* Homo erectus *Skeleton,* Alan Walker and Richard Leakey, eds. (Cambridge, MA: Harvard University Press, 1993).

17. As this book goes to press, the idea that early *Homo erectus* females had narrow birth canals is being challenged by the announcement of what may be the first nearly complete pelvis from an early *Homo erectus* female (Scott W. Simpson et al., "A Female *Homo erectus* pelvis from Gona, Ethiopia," *Science 322* [2008]: 1089–1092).

18. Newborn chimpanzees have an average cranial capacity (a good indicator of brain size) of around 150 cm^3, compared with the approximately 375 cm^3 average for adult chimps, so they are born with about 40 percent of their adult brain size, which is not a bad estimate for the apelike australopithecines. Human newborns average around 350 cm^3, which will reach about 1350–1400 cm^3 by the time they are adults, so the human newborn is born with only about 25 percent of his adult brain size. If newborn cranial capacity in *Homo erectus* split the difference between australopithecines and humans (as the adult capacity of WT 15000 does), it would have been about 33 percent, or one-third, the adult value, which would make it around 300 cm^3. For more detailed information about newborn brain sizes, see Jeremy M. DeSilva and Julie J. Lesnik, "Brain Size at Birth Throughout Human Evolution: A New Method for Estimating Neonatal Brain Size in Homimims," *Journal of Human Evolution,* in press. DeSilva and Lesnik's calculations of neonatal brain sizes are very close to the ones presented here, with the exception of newborn humans, who they estimate are born

with 30% rather than 25% of the average adult brain mass. According to the new report mentioned in the previous note (Simpson et al., 2008), the maximum brain volume of the neonate that the pelvis attributed to a *Homo erectus* female could have birthed would have been 315 cm^3, which is concordant with the 300 cm^3 provided here for WT 15000.

19. In response to the man-the-hunter hypothesis, Adrienne Zihlman has frequently written about what has become known as the woman-the-gatherer hypothesis. See, for example, A. L. Zihlman, "Gathering Stories for Hunting Human Nature: A Review Essay," *Feminist Studies* 11 (1985): 365–377. She and other anthropologists support the idea that women may have been responsible for the earliest inventions of tools, including baby slings. See also S. Linton, "Woman the Gatherer: Male Bias in Anthropology," in *Women in Cross-cultural Perspective,* W. Jacobs, ed. (Champaign–Urbana: University of Illinois Press, 1971).

20. J. M. Adovasio, Olga Soffer, and J. Page, *The Invisible Sex: Uncovering the True Roles of Women in Prehistory* (New York: HarperCollins, 2007).

21. Now if we could just find the relevant fossils! To date, the fossil record is silent on exactly when *Homo erectus* originated, or where (presumably somewhere in Africa, although there are recent murmurings about Asia).

CHAPTER 4: HOW OUR ANCESTORS FOUND THEIR VOICE

1. This hypothesis was first published in Dean Falk, "Prelinguistic Evolution in Early Hominins: Whence Motherese?" *Behavioral and Brain Sciences* 27 (2004): 491–541.

2. Quotation is from Martin Pickford, "Paleoenvironments, Paleoecology, Adaptations, and the Origins of Bipedalism in Hominidae," in *Human Origins and Environmental Backgrounds,* H. Ishida et al., eds. (New York: Springer, 2006), p. 196.

3. Peter deMenocal, "African Climate Change and Faunal Evolution During the Pliocene-Pleistocene," *Earth and Planetary Science Letters* 220 (2004): 3–24.

4. Dean Falk, *Primate Diversity* (New York: W. W. Norton, 2000).

5. Frederick L. Coolidge and Thomas Wynn, "The Effects of the Tree-to-Ground Transition in the Evolution of Cognition in Early *Homo,*" *Before Farming* 2006, no. 4, article 11.

6. E. Balzamo, R. J. Bradley, and J. M. Rhodes, "Sleep Ontogeny in the Chimpanzee; From Two Months to Forty-One Months," *Electroencephalography and Clinical Neurophysiology* 33 (1972): 47–60.

7. B. Carey, "An Active, Purposeful Machine that Comes Out at Night to Play," *New York Times,* October 23, 2007.

8. Michael S. Franklin and Michael J. Zyphur, "The Role of Dreams in the Evolution of the Mind," *Evolutionary Psychology* 3 (2005): 59–79.

9. Zeresenay Alemseged, F. Spoor, W. H. Kimbel, R. Bobe, D. Geraads, D. Reed, and J. G. Wynn, "A Juvenile Early hominin Skeleton from Kikika, Ethiopia," *Nature* 443 (2006): 296–301.

10. In fact, only one other hyoid bone is known from the entire hominin fossil record, and it is from a Neanderthal.

11. Chris P. Sloan, "The Origin of Childhood," *National Geographic* 210 (November 2006): 157.

12. Alemseged et al., p. 156.

13. Ibid., p. 156.

14. According to David Lordkipanidze et al., the range for cranial capacity in Dmanisi is 600–775 cm^3 and the 600 cm^3 capacity is near the mean for another contemporary African species, *Homo habilis*. There is a problem, however, because of the fragmentary nature of the fossils attributed to *Homo habilis*, which some workers think should be included in *Australopithecus*. See David Lordkipanidze et al., "Postcranial Evidence from Early *Homo* from Dmanisi, Georgia," *Nature* 449 (2007): 305–310.

15. Ibid.

16. I. J. Wallace, B. Demes, W. L. Jungers, M. Alvero, and A. Su, "The Bipedalism of the Dmanisi Hominins: Pigeon-Toed Early *Homo?*" *American Journal of Physical Anthropology* 136 (2008): 375–378.

17. Susan Larson et al., "*Homo Floresiensis* and the Evolution of the Hominin Shoulder," *Journal of Human Evolution* 53, no. 6 (December 2007): 718–731.

18. Frank W. Marlowe, "Central Place Provisioning: The Hadza as an Example," in *Feeding Ecology in Apes and Other Primates. Ecological, Physical and Behavioral Aspects,* G. Hohmann, M. M. Robbins, and C. Boesch, eds. (Cambridge, UK: Cambridge University Press, 2006).

19. Primatologists call the social pattern of breaking up into various subgroups at times and merging into larger groups at other times "fission-fusion." It is found in chimpanzees, among other primates.

20. Frank W. Marlowe, "Hunter-Gatherers and Human Variation," *Evolutionary Anthropology* 14 (2005): 54–67.

21. Marlowe, "Central Place Provisioning."

22. Marlowe notes that the interbirth interval for chimpanzees averages 5.6 years, which is not quite twice as long as the average for human foragers. Interbirth interval decreased during hominin evolution, probably partly because of better nutrition (and its associated increased fertility).

23. Marlowe, "Central Place Provisioning," p. 370.

24. Alyssa N. Crittenden and Frank W. Marlowe, "Allomaternal Care Among the Hadza of Tanzania," *Human Nature—An Interdisciplinary Biosocial Perspective* 19, no. 3 (September 2008): 249–262.

25. Kristen Hawkes, "Grandmothers and the Evolution of Human Longevity," *American Journal of Human Biology* 15 (2003): 380–400.

26. Because weaning promotes the return of maternal fertility, the earlier weanings permitted by special weaning foods would also have contributed to an overall increase in group fertility.

27. Sarah Blaffer Hrdy, *Mother Nature: Maternal Instincts and How They Shape the Human Species* (New York: Ballantine, 1999), p. 199.

28. Marlowe, "Central Place Provisioning," p. 363.

CHAPTER 5: THE SEEDS OF LANGUAGE

1. Laurel J. Trainor, C. M. Austin, and R. N. Desjardins, "Is Infant-Directed Speech Prosody a Result of the Vocal Expression of Emotion?" *Psychological Science* 3 (2000): 188–195.

2. R. P. Cooper, J. Abraham, S. Berman, and M. Staska, "The Development of Infants' Preference for Motherese," *Infant Behavior and Development* 20 (1997): 477–488.

3. D. N. Stern, S. Speiker, R. K. Barnett, and K. MacKain, "The Prosody of Maternal Speech: Infant Age and Context Related Changes," *Journal of Child Language* 10 (1983): 1-15.

4. Kyra Karmiloff and Annette Karmiloff-Smith, *Pathways to Language: From Fetus to Adolescent* (Boston: Harvard University Press, 2001), p. 47.

5. Prosodic input to human speech comes mainly from the right sides of their brains, whereas speech itself is generated mostly on the left. Although we will not discuss the neurological substrates of language until Chapter 9, it is worth noting here that, rather than being totally separate from language, as some linguists have argued, tone of voice contributes an extremely important component to language and was probably even more crucial for communication among our earliest nonverbal ancestors. Instead of ignoring it, then, prosody should be analyzed for the evolutionary secrets it can reveal—and tone of voice is nothing if not musical.

6. I was amused when my agent, Deirdre Mullane, told me she had read the proposal for this book while drinking coffee in a public establishment. Apparently, a mother was engaging in motherese with a small baby at the next table just as Mullane was reading the relevant part of the proposal. She said it was like having a live illustrator.

7. Sandra E. Trehub, Laurel J. Trainor, and Anna M. Unyk, "Music and Speech Processing in the First Year of Life," *Advances in Child Development* 24 (1993): 1–35. The authors note that soothing songs are sung less in typical North American and European contexts in which caregivers often withdraw from their infants' rooms before they fall asleep.

8. Anne Fernald, "Human Maternal Vocalizations to Infants as Biologically Relevant Signals: An Evolutionary Perspective," in *Language Acquisition Core Readings,* P. Bloom, ed. (Cambridge, MA: MIT Press, 1994).

9. Marilee Monnot, "Function of Infant-Directed Speech," *Human Nature: An Interdisciplinary Biosocial Perspective* 10 (1999): 415–443; Fernald, "Human Maternal Vocalizations."

10. This kind of thinking is apparent in other aspects of anthropology, such as the contentious issue of how best to interpret human variation. Early explorers who sailed to the Americas noticed that people looked very different from the folks back home, which contributed to the categorization of humans into different races. Contemporary anthropologists such as C. Loring Brace, however, have a different view. Brace observes that for most anatomical features, human variation changes gradually over large distances (this is called clinal variation). For example, as one travels east across Europe, the frequency of the gene for type B blood gradually increases. When human variation is studied in all its complexity, it is clear that the subtleties of gradual change in gene frequencies across large geographical ranges were lost on explorers who plopped themselves down across huge expanses of water. With language, motherese is on one side of the ocean and language is on the other. Linguists who think there is no relationship between the two fail to see links that would become apparent were they on dry land.

11. For details about this "head-turn preference procedure" and other methods see Karmiloff and Karmiloff-Smith, *Pathways to Language.*

12. Karmiloff and Karmiloff-Smith, *Pathways to Language,* p. 44.

13. Ibid., pp. 1–2.

14. If I had it to do over again, which as a grandmother of six I do not, I would talk to, pat, sing to, and play music for my pregnant abdomen!

15. Peter Ladefoged, *Vowels and Consonants: An Introduction to the Sounds of Language,* 2nd ed. (Oxford, UK: Blackwell, 2004).

16. Greg Miller, "Listen, Baby," *Science* 306 (2004): 1127.

17. J. F. Werker and R. C. Tees, "Cross-Language Speech Perception: Evidence for Perceptual Reorganization During the First Year of Life," *Infant Behavioral Development* 7 (1984): 49–63.

18. Patricia K. Kuhl, "A New View of Language Acquisition," *Proceedings of the National Academy of Science USA* 97 (2000): 11860–11857. According to Kuhl's native language neural commitment (NLNC) hypothesis, "Language learning produces dedicated neural networks that code the patterns of native-language speech. The hypothesis focuses on the aspects of language learned early—the statistical and prosodic regularities in language input that lead to phonetic and word learning—and how they influence the brain's future ability to learn language. According to the theory, neural commitment to the statistical and prosodic regularities of one's native language promotes the future use of these learned patterns in higher-order native-language computations. At the same time, NLNC interferes with the processing of foreign-language patterns that do not conform to those already learned." Patricia K. Kuhl, "Early Language Acquisition: Cracking the Speech Code," *Nature Reviews Neuroscience* 5 (2004): 838.

19. Kuhl, "Early Language Acquisition," p. 834.

20. E. D. Thiessen and J. R. Saffran, "Learning to Learn: Infants' Acquisition of Stress-Based Strategies for Word Segmentation," *Language Learning and Development* 3 (2007): 73–100.

21. P. W. Jusczyk, D. M. Houston, and M. Newsome, "The Beginnings of Word Segmentation in English-Learning Infants," *Cognitive Psychology* 39 (1999): 159–207.

22. Miller, "Listen, Baby"; F. M. Tsao, H. M. Liu, and Patricia K. Kuhl, "Speech Perception in Infancy Predicts Language Development in the Second Year of Life: A Longitudinal Study," *Child Development* 75 (2004): 1067–1084.

23. H. M. Liu, Patricia K. Kuhl, and F. M. Tsao, "An Association Between Mothers' Speech Clarity and Infants' Speech Discrimination Skills," *Developmental Science* 6 (2003): F1–F10.

24. Patricia K. Kuhl, S. Coffey-Corina, D. Padden, and G. Dawson, "Links Between Social and Linguistic Processing of Speech in Preschool Children with Autism: Behavioral and Electrophysiological Measures," *Developmental Science* 8, no. 1 (2005): F1–F12.

25. Ibid., p. F10.

26. D. Burnham, C. Kitamura, and U. Vollmer-Conna, "What's New Pussycat? On Talking to Babies and Animals," *Science* 296 (2002): 1435.

27. See J. E. Andruski, Patricia K. Kuhl, and A. Hayashi, "Point Vowels in Japanese Mothers' Speech to Infants and Adults," *Journal of the Acoustical Society of America* 105 (1999): 1095–1096; Burnham, Kitamura, and Vollmer-Conna, "What's New Pussycat?"; Patricia K. Kuhl, J. E. Andruski, J. A. Chistovich, L. A. Chistovich, E. V. Kozhevnikova, V. L. Ryskina, E. I. Stoljarova, U. Sundberg, and F. Lacerda, "Cross-Language Analysis of Phonetic Units in Language Addressed to Infants," *Science* 277 (1997): 684–686.

28. Liu, Kuhl, and Tsao, "An Association Between Mothers' Speech Clarity."

29. B. deBoer and Patricia K. Kuhl, "Investigating the Role of Infant-Directed Speech with a Computer Model," *Acoustics Research Letters Online* 4, no. 4 (2003): 129–134.

30. K. Hirsh-Pasek and R. Treiman, "Doggerel; Motherese in a New Context," *Journal of Child Language* 9 (1982): 229–237.

31. Such statements are said to be deictic, an adjective related to "index" that means demonstrating by pointing or other forms of reference. One often points with the index finger when making such statements.

32. Andrew N. Meltzoff, "'Like Me': A Foundation for Social Cognition," *Developmental Science* 10 (2007): 126–134; Andrew N. Meltzoff, "Imitation, Objects, Tools and the Rudiments of Language in Human Ontogeny," *Human Evolution* 3 (1988): 45–64.

33. Andrew N. Meltzoff and J. Decety, "What Imitation Tells Us About Social Cognition: A Rapprochement Between Developmental Psychology and Cognitive Neuroscience," *Philosophical Transactions of the Royal Society of London,* B 358 (2003): 491–500.

34. So-called mirror neurons that are discussed in Chapter 9 come into play here.

35. Meltzoff and Decety, "What Imitation Tells Us About Social Cognition," p. 494.

36. Patricia K. Kuhl and Andrew N. Meltzoff, "Speech as Intermodal Object of Perception," in *Perceptual Development in Infancy: The Minnesota Symposium on Child Phonology,* A. Yonas, ed. (Hillsdale, NJ: Lawrence Erlbaum and Associates, 1988).

37. S. J. Cowley, S. Moodley, and A. Fiori-Cowley, "Grounding Signs of Culture: Primary Intersubjectivity in Social Semiosis," *Mind, Culture and Activity* 11 (2004): 109–132.

38. Karmiloff and Karmiloff-Smith, *Pathways to Language.*

39. Thiessen and Saffran, "Learning to Learn."

40. When you think about it, it's pretty amazing that fetuses begin to distinguish the intonations and rhythms of speech before they can see the sources of the sounds they hear. Newborns, on the other hand, work at perceiving visual as well as auditory patterns, which anyone with access to a baby can verify. Try, for example, placing an infant in front of a window that has light streaming through lace curtains. The baby will stare at the scene in apparent fascination, sharpening her three-dimensional form and depth perception in the process. She is learning to see. Newborns also apply the pattern-processing skills they honed on prenatal listening to the task of developing and integrating information from hearing, seeing, feeling, and movement. No small job, that.

41. This revelation is due largely to the hard work of German researchers Kathleen Wermke of the University of Wuerzburg, Werner Mende of the Berlin-Brandenburg Academy of Science, and their colleagues.

42. Personal communication to author in letter from Werner Mende and Kathleen Wermke dated September 26, 2006, quoted with permission.

43. As such, infants' early cries appear to manifest the recursion that linguists worry so much about. W. T. Fitch, M. D. Hauser, and N. Chomsky, "The Evolution of the Language Faculty: Clarifications and Implications," *Cognition* 97 (2005): 179–210.

44. Kathleen Wermke, Werner Mende, C. Manfredi, and P. Bruscaglioni, "Developmental Aspects of Infants' Cry Melody and Formants," *Medical Engineering & Physics* 24 (2002): 501–514.

45. Kathleen Wermke and Werner Mende, "Melody as a Primordial Legacy from Early Roots of Language," *Behavioral and Brain Sciences* 29 (2006): 300.

46. Wermke, Mende, Manfredi, and Bruscaglioni, "Developmental Aspects of Infants' Cry Melody and Formants."

47. So my grandmother was not that far off the mark after all when she claimed that babies cry to "exercise their little lungs"—she just didn't include enough of their anatomy in her assertion!

48. Kathleen Wermke, Daniel Leising, and Angelika Stellzig-Eisenhauer, "Relation of Melody Complexity in Infants' Cries to Language Outcome in the Second Year of Life: A Longitudinal Study," *Clinical Phonetics & Linguistics* 21, no. 11-12 (2007): 961–973.

49. Ibid.

50. Siobhan Holowka and Laura Ann Petitto ("Left Hemisphere Cerebral Specialization for Babies While Babbling," *Science* 297 [2002]: 1515) define babbles as vocalizations that contain a reduced set of a language's possible sounds (phonetic units), have reduplicated (repeated) syllabic organization (consonant-vowel alternations), and are produced without apparent meaning.

51. A. Levitt and J. G. Aydelott Utman, "From Babbling Towards the Sound Systems of English and French: A Longitudinal Two-Case Study," *Journal of Child Language* 19 (1992): 19–49.

52. T. Imada, Y. Zhang, M. Cheour, S. Taulu, A. Ahonen, and Patricia K. Kuhl, "Infant Speech Perception Activates Broca's Area: A Developmental Magnetoencephalography Study," *Neuroreport* 17 (2006): 957–962; Holowka and Petitto, "Left Hemisphere Cerebral Specialization."

53. Charles A. Ferguson, "Talking to Children: A Search for Universals," in *Universals of Human Language,* J. H. Greenberg, ed. (Stanford: Stanford University Press, 1978).

54. According to Ferguson, "such modifications have an innate basis in pan-human child-care behaviors, but the details in every speech community are largely conventionalized (i.e., culturally shaped) and in part arise directly from interactional needs and imitation of children's behavior. The modificational features are variable in incidence, but constitute a surprisingly cohesive set of linguistic features that may be regarded as a 'register' in the language user's repertoire. This 'baby talk' register is an important factor in the socialization of children, apparently assisting in the acquisition of linguistic structure, the

development of interactional patterns, the transmission of cultural values, and the expression of the user's affective relationship with the addressees." Ferguson, "Talking to Children," p. 215.

55. Elinor Ochs, "Indexing Gender," in *Rethinking Context: Language as an Interactive Phenomenon,* A. Duranti and C. Goodwin, eds. (Cambridge, UK: Cambridge University Press, 1992), p. 349.

56. Elinor Ochs and Bambi B. Schieffelin, "Language Acquisition and Socialization: Three Developmental Stories and Their Implications," in *Culture Theory: Essays on Mind, Self, and Emotion,* R. A. Shweder and R. A. LeVine, eds. (Cambridge, UK: Cambridge University Press, 1984), p. 293.

57. Bambi B. Schieffelin, *The Give and Take of Everyday Life: Language Socialization of Kaluli Children* (Cambridge, UK: Cambridge University Press, 1990), p. 70.

58. Ibid., p. 71.

59. Ibid., p. 71.

60. Ochs and Schieffelin, "Language Acquisition and Socialization," p. 291.

61. Ibid., p. 291.

62. Ibid., p. 291.

63. Ibid., p. 292.

64. Ibid., p. 292.

65. Ibid., p. 302.

66. Elinor Ochs, "Talking to Children in Western Samoa," *Language in Society* 14 (1982): 89.

67. Ochs and Schieffelin, "Language Acquisition and Socialization," pp. 295–296.

68. Ibid., p. 297.

69. Shirley Brice Heath, *Ways With Words: Language, Life, and Work in Communities and Classrooms* (Cambridge, UK: Cambridge University Press, 1983).

70. Monique Tenette Mills, "Phonological Features of African-American Vernacular English in Child-Directed Versus Adult-Directed Speech" (master's thesis, Ohio State University, 2004).

71. Clifton Pye, "Quiché Mayan Speech to Children," *Journal of Child Language* 13 (1986): 85–100.

72. N. B. Ratner and Clifton Pye, "Higher Pitch in BT is Not Universal: Acoustic Evidence from Quiché Mayan," *Journal of Child Language* 11 (1984): 521.

73. Ibid., p. 520.

74. Clifton Pye, "The Acquisition of K'iché Maya," in *The Crosslinguistic Study of Language Acquisition,* vol. 3, D. I. Slobin, ed. (Hillsdale, NJ: Lawrence Erlbaum and Associates, 1991), p. 244.

CHAPTER 6: WHAT'S IN A NAME?

1. Ellen Dissanayake, "Antecedents of the temporal arts in early mother-infant interaction," in *The Origins of Music,* N. L. Wallin, B. Merker, and S. Brown, eds. (Cambridge, MA: MIT Press, 2000), p. 391. In 1997, Dissanayake, of the University of Washington, gave a talk about ancestral mother-infant interactions and their role in the emergence of the temporal arts, at an interna-

tional workshop on biomusicology that we both attended, which sparked my interest in motherese.

2. Dean Falk, "Prelinguistic Evolution in Early Hominins: Whence Motherese?" *Behavioral and Brain Sciences* 27 (2004): 491–503.

3. Ibid.

4. D. Bickerton, "Mothering Plus Vocalization Doesn't Equal Language," *Behavioral and Brain Sciences* 27 (2004): 505.

5. Anderson is a cultural anthropologist who lived with the Sami (formerly called Laplanders) for five years, during which she became fluent in their language (which is related to Finnish). She has taught anthropological linguistics (and other subjects) for many years at Purdue University.

6. D. L. Everett, "Cultural Constraints on Grammar and Cognition in Piraha," *Current Anthropology* 46 (2005): 621–646.

7. Steven Pinker and Ray Jackendoff, "The Faculty of Language: What's Special About It?" *Cognition* 95 (2005): 202, 210.

8. Paul Bloom, *How Children Learn the Meanings of Words* (Cambridge, MA: MIT Press, 2000).

9. Paul Bloom, "Is Grammar Special?" *Current Biology* 9 (1999): R1127–R128. Bloom, a linguist at Yale University, thinks at least two mechanisms are involved in learning words. The first applies to learning concrete nouns and verbs, such as "dog" and "smile," which Bloom suggests relies on the same systems of inference and memory that apply when children acquire social knowledge. The second learning mechanism entails grammatical cues and is used to learn more abstract nouns and verbs, such as "story" and "think."

10. Pinker and Jackendoff, "The Faculty of Language," p. 211.

11. Kyra Karmiloff and Annette Karmiloff-Smith, *Pathways to Language: From Fetus to Adolescent* (Boston: Harvard University Press, 2001), p. 57.

12. Pinker and Jackendoff, "The Faculty of Language."

13. Karmiloff and Karmiloff-Smith, *Pathways to Language*, p. 132.

14. G. Miller, "Listen, Baby," *Science* 306 (2004): 1127; F. M. Tsao, H. M. Liu, and Patricia K. Kuhl, "Speech Perception in Infancy Predicts Language Development in the Second Year of Life: A Longitudinal Study," *Child Development* 75 (2004): 1067–1084.

15. Jinyun Ke, J. Minett, C. P. Au, and S. Y. Wang, "Self-Organization and Selection in the Emergence of Vocabulary," *Complexity* 7 (2002): 41–54.

16. J. Ganger and M. R. Brent, "Reexamining the Vocabulary Spurt," *Developmental Psychology* 40 (2004): 631.

17. D. Poulin-Dubois, S. Graham, and L. Sippola, "Early Lexical Development: The Contribution of Parent Labeling and Infants' Categorization Abilities," *Journal of Child Language* 22 (1995): 325–343.

18. Ganger and Brent, "Reexamining the Vocabulary Spurt."

19. B. A. Goldfield and J. S. Reznick, "Early Lexical Acquisition: Rate, Content, and the Vocabulary Spurt," *Journal of Child Language* 17 (1990): 171–183.

20. L. Naigles and E. Hoff-Ginsberg, "Why Are Some Verbs Learned Before Other Verbs? Effects on Input Frequency and Structure on Children's Early Verb Use," *Journal of Child Language* 25 (1998): 95–120.

21. A. Fernald and N. Hurtado, "Names in Frames: Infants Interpret Words in Sentence Frames Faster than Words in Isolation," *Developmental Science* 9 (2006): F33–F40.

22. T. Cameron-Faulkner, E. Lieven, and M. Tomasello, "A Construction-Based Analysis of Child-Directed Speech," *Cognitive Science* 27 (2003): 843–873.

23. Karmiloff and Karmiloff-Smith, *Pathways to Language*, p. 62.

24. E. L. Bavin, "Language Acquisition in Crosslinguistic Perspective," *Annual Review of Anthropology* 24 (1995): 378.

25. Karmiloff and Karmiloff-Smith, *Pathways to Language*.

26. Ibid., p. 63.

27. Ibid., p. 77.

28. Ibid., p. 132.

29. M. D. S. Braine, "Children's First Word Combinations," *Monographs of the Society for Research in Child Development* 41 (1976), serial no. 164.

30. R. M. Golinkoff, K. Hirsh-Pasek, K. M. Cauley, and L. Gordon, "The Eyes Have It: Lexical and Syntactic Comprehension in a New Paradigm," *Journal of Child Language* 14 (1987): 23–46; Karmiloff and Karmiloff-Smith, *Pathways to Language,* pp. 91–92.

31. Karmiloff and Karmiloff-Smith, *Pathways to Language,* p. 115.

32. R. A. Brown, *A First Language: The Early Stages* (Cambridge, MA: Harvard University Press, 1973).

33. Karmiloff and Karmiloff-Smith, *Pathways to Language,* pp. 95–96.

34. M. J. Farrar, "Discourse and the Acquisition of Grammatical Morphemes," *Journal of Child Language* 17 (1990): 607–624.

35. V. Kempe and P. Brooks, "The Role of Diminutives in the Acquisition of Russian Gender: Can Elements of Child-Directed Speech Aid in Learning Morphology?" *Language Learning* 51 (2001): 221–256.

36. The degree to which people are born with an innate sense of grammar is a controversial topic that lies at the heart of the argument over whether motherese was pivotal for the origins of language. On one side of the debate are the followers of linguist Noam Chomsky, who believe that a universal grammar underlies all language and is hardwired into humans. Decades ago (as outlined in his book *Knowledge of Language: Its Nature, Origin, and Use* [New York: Praeger, 1986]), Chomsky formulated the now-famous idea that humans are born with an innate grasp of "Universal Grammar" that constrains how words are created, combined, and moved within sentences to form questions, relative clauses, and so on. In theory, it is their inborn grammar that allows infants to learn the particulars of their native languages and to begin generating proper utterances with ease and fluidity. This idea, which still has a strong foothold in linguistics literature, is probably behind linguists' reluctance to accept the importance of motherese for infants' development of grammar and language.

In contrast to Chomsky, Jackendoff and Pinker's ideas about how infants learn language focus on unusual sayings and idioms. (See their 2005 article, "The Nature of the Language Faculty and Its Implications for the Evolution of Language" [Reply to Fitch, Hauser, and Chomsky], *Cognition* 97: 211–225.)

Most of these sayings have proper syntax, such as "son of a gun" (a noun phrase), "down in the dumps" (a verb phrase), and "the jig is up" (a sentence), while a few, such as "by and large" and "far be it from me" do not. Jackendoff and Pinker observe that English (and presumably other languages) has many "syntactic nuts" and, further, the number of sayings that speakers know may even approach the number of words in their vocabularies. Like whole words and their grammatical parts, these sayings are stored in memory. Although other linguists have more or less ignored idioms and other odd constructions, Jackendoff and Pinker conclude that the rules for language reside in mentally stored sayings rather than in an innate grasp of universal grammar: "The 'construction-based' view of language that emerges from these considerations, if correct, has consequences for the study of language processing, acquisition, and, most germane to the present discussion, evolution. If a speaker's knowledge of language embraces all the words, all the constructions, and all the general rules, coded in the same format, then there is no coherent subset of language that 'delineate[s] an abstract core of computational operations.'" (Quoted from p. 222 of their article.)

One needs a good memory to store a large number of idioms and constructions that are odd but, for the most part, "commandeer the same computational machinery" used to combine words into proper speech. This, in turn, suggests that the increase in brain size that occurred during our ancestors' evolution may have been associated, in part, with a sharpened capacity for storing and retrieving memories, including linguistic ones.

37. B. L. Finlay and R. B. Darlington, "Linked Regularities in the Development and Evolution of Mammalian Brains," *Science* 268 (1995): 1578–1584.

38. C. S. Goodman and B. C. Coughlin, "The Evolution of Evo-Devo Biology," *Proceedings of the National Academy of Sciences USA* 97 (2000): 4424–4425.

39. To quote Wallace Arthur of the National University of Ireland, "Haeckel's idea of recapitulation—the fleeting occurrence in embryos of 'advanced' animals of certain features of their ancestors—is not dead, despite many reports to the contrary." Arthur, "The Search for Novelty," *Nature* 447 (2007): 261–262.

40. Although Robbins Burling of the University of Michigan agrees that tone of voice evolved from primate calls, he also thinks that it "amounts to an invasion of language by something that is fundamentally different." (Robbins Burling, "Primate Calls, Human Language, and Nonverbal Communication," *Current Anthropology* 34 [1993]: 30.) Burling parts ways, however, with other linguists who think that language emerged suddenly, when he states, "I do believe that language emerged very gradually from something other than a primate call system." (See Robbins Burling, "Prosody Does Not Equal Language," *Behavioral and Brain Sciences* 27 [2004]: 509.) Nevertheless, Burling doubts that motherese was important for the invention of speech sounds, words, or rules for forming sentences.

Derek Bickerton, a linguist from the University of Hawaii, has a different view. Bickerton envisions a protolanguage that evolved between two million and three million years ago in the context of scavenging for food. Such a lifestyle, he writes, "virtually obligated a foraging pattern in which groups split into smaller units for search procedures but had to regroup to exploit substantial finds (such as the carcasses of large thick-skinned megafauna, which hominids' possession of stone cutting tools would have enabled them to exploit in advance of other scavengers). This pattern required each sub-group to be able to inform the other sub-groups what it had found, in order to secure the best disposition of resources. In such a context, the smallest handful of single-unit utterances would have paid off in terms of survival" (Derek Bickerton, "Language Evolution: A Brief Guide for Linguists," *Lingua* 117: 515).

Further, Derek believes that protolanguage consisted of culturally based "symbolic, referential units" (words) in the form of nouns and verbs, but that it lacked grammatical structure. He also likens it to the "telegraphic speech produced, before age two, by children" (which is as close as he comes to including women and children in his scenario) (ibid, p. 516). Bickerton sees no intermediate stage, however, between such protolanguage two million to three million years ago and the earliest true language with syntax, which he suggests occurred between 140,000 and 90,000 years ago in conjunction with the appearance of *Homo sapiens*. Thus, unlike Burling, Bickerton appears to view the emergence of full-blown language as abrupt rather than gradual. Whether they think language happened gradually or abruptly, theorists such as Burling and Bickerton are adamant that language is too different from primate calls to have evolved from them by means of natural selection. Instead, such theorists often suggest that humanlike language entailed changes in the brain associated with one or more genetic mutations that occurred by chance in a small population of *Homo sapiens*. Although this explanation might appear satisfying at first, it is frustratingly vague and begs the very question that linguists have asked: How exactly did the sounds and syntax of speech begin in the first place, especially if they appeared virtually overnight (geologically speaking) during the course of human evolution?

41. Problems arise when different scientists cut the linguistic pie differently. For example, in what is perceived as a marked departure from his earlier work (Pinker and Jackendoff, "The Faculty of Language: What's Special About It?"), Chomsky and his colleagues currently divide human language into two parts (M. D. Hauser, N. Chomsky, and W. T. Fitch, "The Faculty of Language: What It Is, Who Has It, and How Did It Evolve?" *Science* 298 (2002): 1569–1579). One is a broad faculty of language, which includes features shared with other animals, such as an ability to discriminate speech sounds or to imitate vocalizations (e.g., parrots). The second part is a narrow faculty of language, which is asserted to be recently evolved and unique to humans. And narrow it is because it represents the only mechanism of language that Chomsky (now) suggests is unique to humans, namely an ability to embed something in something else of the same type, called recursion, which creates an open-ended and limitless system for communicating. As defined by Pinker and Jackendoff: "Recursion consists of embedding a constituent in a constituent of the same type, for example, a relative clause inside a relative clause (*a book that was written by the novelist you met last night*), which automatically confers the ability to do so ad libitum (e.g., *a book [that was written by the novelist [you met on the night [that we decided to buy the boat [that you liked so much]]]]*). This does not exist in phonological

structure: a syllable, for instance, cannot be embedded in another syllable" (quoted from p. 208 of their article).

Since Hauser et al. published their views on recursion being unique to humans, another group has demonstrated that "European starlings accurately recognize acoustic patterns defined by recursive, self-embedding, context-free grammar. . . . Thus, the capacity to classify sequences from recursive, centre-embedded grammars is not uniquely human." T. Q. Gentner, K. M. Fenn, D. Margoliash, and H. C. Nusbaum, "Recursive Syntactic Pattern Learning by Songbirds," *Nature* 440 (2006): 1204.

42. Jackendoff and Pinker, "The Nature of the Language Faculty and Its Implications for the Evolution of Language."

43. Thus, according to Jackendoff and Pinker, "Only after these more basic aspects of linguistic communication are in place could there be any adaptive advantage to the system's developing regimented syntactic means to arrange words into larger utterances, so that semantic relations among words could be expressed in conventionalized fashion." Ibid., p. 223. Despite being linguists, Pinker and Jackendoff even entertain the possibility that language was directly targeted by natural selection and, as such, represents an adaptation for communicating knowledge and intentions. If their "construction-based" view of language evolution is right, the evolution of big brains with presumably big memories may have been important for banking the idioms and expressions that contain the kernels of syntax.

44. M. D. Hauser, *The Evolution of Communication* (Cambridge, MA: MIT Press, 1996).

45. D. L. Cheney and R. M. Seyfarth, *How Monkeys See the World* (Chicago: University of Chicago Press, 1990).

46. Ke, Minett, Au, and Wang, "Self-Organization and Selection in the Emergence of Vocabulary."

47. A. Fernald and H. Morikawa, "Common Themes and Cultural Variations in Japanese and American Mothers' Speech to Infants," *Child Development* 64 (1993): 637–656.

48. S. Kojima, *A Search for the Origins of Human Speech* (Kyoto, Japan: Kyoto University Press, 2003).

49. P. J. Horne and F. Lowe, "On the Origins of Naming and Other Symbolic Behavior," *Journal of Experimental Analysis of Behavior* 65 (1996): 185–241.

50. S. Harnad, "The Origin of Words: A Psychophysical Hypothesis," in *Communicating Meaning, The Evolution and Development of Language,* B. M. Velichkovsky and D.M. Rumbaugh, eds. (Hillsdale, NJ: Lawrence Erlbaum and Associates, 1996).

51. Steven Mithen, *The Singing Neanderthals* (Cambridge, MA: Harvard University Press, 2006), pp. 203–204. Mithen is a professor of early prehistory at Reading University in England.

52. Falk, "Prelinguistic Evolution in Early Hominins," p. 502–503.

53. Peter F. MacNeilage, "The Frame/Content Theory of Evolution of Speech Production," *Behavioral and Brain Sciences* 21 (1998): 499–546; MacNeilage, "The Explanation of 'Mama,'" *Behavioral and Brain Sciences* 23 (2000): 440–441. MacNeilage is from the University of Texas at Austin.

54. H. I. Goldman, "Parental Reports of 'MAMA' Sounds in Infants: An Exploratory Study," *Journal of Child Language* 28 (2001): 497–506.

55. R. Tincoff and P. W. Jusczyk, "Some Beginnings of Word Comprehension in Six-Month-Olds," *Psychological Science* 10 (1999): 172–175.

56. Cognitive neuroscientist Merlin Donald defines protolanguage as "a prototype, an approximation of the finished product that contains some of the essential features of language, and that could later have evolved into full-fledged language." See M. Donald, "Preconditions for the Evolution of Protolanguages," in *The Descent of Mind: Psychological Perspectives on Hominid Evolution,* M. C. Corballis and I. Lea, eds. (Oxford: Oxford University Press, 1999).

57. M. Kawai, "Newly Acquired Pre-Cultural Behavior of the Natural Troop of Japanese Monkeys on Koshima Islet," *Primates* 6 (1965): 1–30.

58. M. Nakamichi, E. Kato, Y. Kojima, and N. Itoigawa, "Carrying and Washing of Grass Roots by Free-Ranging Japanese Macaques at Katsuyama," *Folia Primatologica* 69 (1998): 35–40.

59. Kawai, "Newly Acquired Pre-Cultural Behavior," pp. 5 and 8.

60. C. Boesch and H. Boesch-Achermann, "Dim Forest, Bright Chimps," *Natural History* 100 (1991): 50–57.

61. J. D. Pruetz and P. Bertolani, "Savanna Chimpanzees, *Pan troglodytes Verus,* Hunt with Tools," *Current Biology* 17 (2007): 412–417.

62. Ke, Minett, Au, and Wang, "Self-Organization and Selection in the Emergence of Vocabulary."

63. S. Kirby, *Function, Selection and Innateness: The Emergence of Language Universals* (New York: Oxford University Press, 1999). What Kirby describes is an example of a spontaneous process called self-organization, which characterizes many dynamic systems.

CHAPTER 7: SHE SHALL HAVE MUSIC

1. Stephen Jay Gould and R. Lewontin, "The Spandrels of San Marco and the Panglossion Paradigm: A Critique of the Adaptationist Programme," *Proceedings of the Royal Society of London B* 205 (1979): 581–598.

2. Steven Pinker, *How the Mind Works* (New York: W. W. Norton, 1997), pp. 534–538.

3. Charles Darwin, *The Descent of Man, and Selection in Relation to Sex,* 2 vols. (London: John Murray, 1890), p. 572.

4. So which has more credibility—music as an evolutionary appetizer that preceded language, or as a dessert that followed? The answer depends on how music is defined and whether scholars take a top-down or bottom-up approach. If one begins from the vantage point of full-blown music and works backward, then it is difficult to see how a piece such as Vivaldi's *Four Seasons,* for example, could have evolved from something like the relatively simple calls of chimpanzees. It is also hard to imagine specific functions that would have caused such music to be a direct target of natural selection in the past, despite it being intensely pleasurable to listen to today. Cheesecake, indeed! But if so, then so are the sounds and images evoked by poetry, such as these lines from Elizabeth Barrett Browning's sonnet "How Do I Love Thee?":

How do I love thee? Let me count the ways.
I love thee to the depth and breadth and height
My soul can reach, when feeling out of sight
For the ends of being and ideal grace.

Or Macbeth's nihilistic soliloquy upon hearing about the death of
　Lady Macbeth (William Shakespeare, *Macbeth,* V, v, 19–28):

Tomorrow, and tomorrow, and tomorrow,
Creeps in this petty pace from day to day
To the last syllable of recorded time,
And all our yesterdays have lighted fools
The way to dusty death. Out, out, brief candle!
Life's but a walking shadow, a poor player
That struts and frets his hour upon the stage
And then is heard no more: it is a tale
Told by an idiot, full of sound and fury,
Signifying nothing.

If music is auditory cheesecake, then surely such sweet sounds of language
should also be thought of as a kind of evolutionary dessert in that they emerged
relatively recently and were therefore "saved for last." For further discussion of
the various views, see Steven Mithen, *The Singing Neanderthals* (Cambridge,
MA: Harvard University Press, 2006).

　5. Bruno Nettl, "An Ethnomusicologist Contemplates Universals in Musical Sound and Musical Culture," in *The Origins of Music,* N. L. Wallin, B. Merker,
and S. Brown, eds. (Cambridge, MA: MIT Press, 2000), p. 406.

　6. Nettl also speculates that:

Most societies have in their music, either as the main style but more commonly as a special repertory, something I might label (cringing because musical
complexity is not easily measured and subject to biases brought about by a culture that worships complexity) as "the world's simplest style." It consists of songs
that have a short phrase repeated several or many times, with minor variations,
using three or four pitches within a range of a fifth. This kind of music is interestingly widespread. . . . It appears to have been the only style, or the principal
style, of some peoples living in widely separated isolated areas of the world. In
addition, it is found in societies whose music is otherwise more complex, and
here it is often relegated to the accompaniment of children's games, to games
generally, and to obsolete rituals. We have reason to believe that it is old mate-
rial, associated as it is with social contexts once central to the culture but over-
taken by more complex music.

　Nettl, "An Ethnomusicologist Contemplates Universals," pp. 468–469.

　7. Sandra E. Trehub, Laurel J. Trainor, and A. M. Unyk, "Music and Speech
Processing in the First Year of Life," *Advances in Child Development and Behavior*
24 (1993): 3–4.

8. Laurel J. Trainor, C. M. Austin, and R. N. Desjardins, "Is Infant-Directed Speech Prosody a Result of the Vocal Expression of Emotion?" *Psychological Science* 11 (2000): 188–195.

9. Ellen Dissanayake, "If Music Is the Food of Love, What About Survival and Reproductive Success?" *Musicae Scientiae* Special Issue (2008): 169.

10. Mithen, *The Singing Neanderthals*, p. 15.

11. Jaak Panksepp and Günther Bernatzky, "Emotional Sounds and the Brain: The Neuro-Affective Foundations of Musical Appreciation," *Behavioural Processes* 60 (2002): 133–155.

12. Trehub, Trainor, and Unyk, "Music and Speech Processing in the First Year of Life."

13. Aniruddh D. Patel, "Language, Music, Syntax and the Brain," *Nature Neuroscience* 6 (2003): 674–681.

14. As Patel discusses, however, the neurological overlap in syntactical processing between language and music is only partial. See B. Maess, S. Koelsch, T. C. Gunter, and A. D. Friederici, "Musical Syntax Is Processed in Broca's Area: An MEG Study," *Nature Neuroscience* 4 (2001): 540–545.

15. Aniruddh D. Patel, J. R. Iversen, and J. C. Rosenberg, "Comparing the Rhythm and Melody of Speech and Music: The Case of British English and French," *Journal of the Acoustical Society of America* 119 (2006): 3034–3047.

16. Thus, the musical interval (the distance between two notes) called the major third (C to E) is the basis for Beethoven's glorious "Ode to Joy" in the *Choral Symphony* of his Ninth Symphony, while the minor third (C to E-flat) is prominent in the ominous first movement of his famous "fate-knocking-on-the-door" Fifth Symphony. (See D. Cooke, *The Language of Music* [Oxford: Oxford University Press, 1959]. For discussion of Cooke's and others' work on music and emotions, see Mithen, *The Singing Neanderthals*, pp. 90–92.) These pieces illustrate an important way the *meaning* of music unfolds as the listener experiences patterns of tension and resolution by anticipating and unconsciously tracking the intervals between notes—i.e., according to the musical syntax Patel so eloquently discusses. At a more general level, the patterns of intervals between notes determine, to a large extent, whether music is perceived as dissonant or consonant. This depends on the particular vibrations that take place on a membrane in the cochlea of the inner ear and are then transmitted to the brain, where they undergo further processing.

17. Isabelle Peretz of the University of Montreal and Robert Zatorre of McGill University do research in a special concert hall that is part of the International Laboratory for Brain, Music, and Sound Research. Approximately twenty-five seats in the hall have wireless sensors that monitor listeners' heart rate and skin and facial responses during live performances. They also can record their unfolding reactions on Palm Pilots. Another approach is to use medical imaging technology (such as functional magnetic resonance imaging, or fMRI) to study what's going on in the brains of listeners, and Zatorre and his colleagues have made startling discoveries. For example, the shivers-down-the-spine feeling that is sometimes caused by music is processed deep within the brain in pleasure centers that also respond to food and sex. (For details, see

M. Balter, "Study of Music and the Mind Hits a High Note in Montreal," *Science* 315 [2007]: 758–759.)

18. Robert J. Zatorre, D. W. Perry, C. A. Beckett, C. F. Westbury, and A. C. Evans, "Functional Anatomy of Musical Processing in Listeners with Absolute Pitch and Relative Pitch," *Proceedings of the National Academy of Sciences USA* 95 (1998): 3172–3177.

19. For a general discussion of brain lateralization, see Dean Falk, *Braindance,* revised and expanded edition (Gainesville: University Press of Florida, 2004), pp. 104–109.

20. Thomas Geissmann, "Gibbon Songs and Human Music from an Evolutionary Perspective," in *The Origins of Music,* Wallin, Merker, and Brown, eds.

21. For more information about gibbons and other primates, see Dean Falk, *Primate Diversity* (New York: W. W. Norton, 2000).

22. Each gibbon species has its own particular song, but they share certain elements that were probably present in the song of an earlier ancestor, according to gibbon expert Thomas Geissmann of the University of Zürich-Irchel. Thus, the ancestor of modern gibbons probably produced duets and had long, loud, and complex songs. Their phrases would have consisted of pure tones, and males' contributions would have developed gradually from simpler to more complex phrases. Females probably provided a great call that accelerated moderately and then slowed down at the end, at which point it would have been accompanied by acrobatics. Males may or may not have engaged in similar displays (Geissmann, "Gibbon Songs and Human Music").

23. Tecumseh Fitch, "The Evolution of Music in Comparative Perspective," *Annals of the New York Academy of Science* 1060 (2005): 29–49.

24. H. Koda, C. Oyakawa, A. Kato, Nobuo Masataka, "Experimental Evidence for the Volitional Control of Vocal Production in an Immature Gibbon," *Behaviour* 144 (2007): 681–692.

25. D. R. Leighton, "Gibbons: Territoriality and Monogamy," in *Primate Societies,* B. B. Smuts, C. L. Cheney, R. M. Seyfarth, R. W. Wrangham, and T. T. Struhsaker, eds. (Chicago: University of Chicago Press, 1987), p. 140.

26. Dian Fossey, *Gorillas in the Mist* (Boston: Houghton Mifflin, 1983).

27. George B. Shaller, *The Mountain Gorilla* (Chicago: University of Chicago Press, 1963), p. 223.

28. D. Crockford, I. Herbinger, L. Vigilant, and C. Boesch, "Wild Chimpanzees Produce Group-Specific Calls: A Case for Vocal Learning," *Ethology* 110 (2004): 221–243.

29. Jane Goodall, *The Chimpanzees of Gombe* (Cambridge, MA: Belknap Press of Harvard University Press, 1986).

30. Thomas Geissmann, "Gibbon Songs and Human Music from an Evolutionary Perspective," p. 118. Geissmann refers to G. Ewens as his source for this observation.

31. Ibid., p. 119.

32. J. D. Newman, "Motherese by Any Other Name: Mother-Infant Communication in Non-Hominin Mammals," *Behavioral and Brain Sciences* 27 (2004): 519–520; J. P. Lorberbaum, J. D. Newman, A. R. Horwitz, J. R. Dubno,

R. B. Lydiard, M. B. Hamner, D. E. Bohning, and M. S. George, "A Potential Role for Thalamocingulate Circuitry in Human Maternal Behavior," *Biological Psychiatry* 51 (2002): 431–445.

33. Newman, "Motherese by Any Other Name."

34. Ibid., p. 519.

35. M. Biben, D. Symmes, and D. Bernhards, "Contour Variables in Vocal Communication Between Squirrel Monkey Mothers and Infants," *Developmental Psychobiology* 22 (1989): 617–631.

36. Nobuo Masataka, *The Onset of Language* (Cambridge, UK: Cambridge University Press, 2003), p. 129.

37. Nobuo Masataka, "Music, Evolution and Language," *Developmental Science* 10 (2007): 36.

38. Fitch, "The Evolution of Music in Comparative Perspective," p. 44.

39. Geoffrey F. Miller, "Evolution of Music Through Sexual Selection," in *The Origins of Music,* N. L. Wallin, B. Merker, and S. Brown, eds. (Cambridge, MA: MIT Press, 2000): 329–360; Fitch, "The Evolution of Music in Comparative Perspective"; Sandra E. Trehub, "Human Processing Predispositions and Musical Universals," in *The Origins of Music,* Wallin, Merker, and Brown, eds.

40. Fitch, "The Evolution of Music in Comparative Perspective," p. 42.

41. M. Papoušek, H. Papoušek, and D. Symmes, "The Meaning of Melodies in Motherese in Tone and Stress Languages," *Infant Behavior and Development* 14 (1991): 415–440; Inge Cordes, "Melodic Contours as a Connecting Link Between Primate Communication and Human Singing," presented at 5th ESCOM Conference, September 8–13, 2003, Hanover University of Music and Drama, Germany, *Music Therapy Today* 4, no. 5, available at http://musictherapyworld.net.

42. Peter Marler, "Aggregation and Dispersal: Two Functions in Primate Communication," in *Primates: Studies in Adaptation and Variability,* P. C. Jay, ed. (New York: Holt, Rhinehart and Winston, 1968). See also: G. Tembrock, *Biokommunikation. Informationsübertragung im biologischen Bereich, Teil II* (Berlin: Akademie-Verlag, 1971).

43. Panksepp and Bernatzky, "Emotional Sounds and the Brain."

44. Ibid., p. 143.

45. Trehub, "Human Processing Predispositions and Musical Universals."

46. Sandra E. Trehub, "The Developmental Origins of Musicality," *Nature Neuroscience* 6 (2003): 669–673.

47. A. Volkova, Sandra E. Trehub, E. G. and Schellenberg, "Infants' Memory for Musical Performances," *Developmental Science* 9 (2006): 583–589.

48. Trehub, "The Developmental Origins of Musicality," p. 670.

49. Laurel J. Trainor, E. D. Clark, A. Huntley, and B. A. Adams, "The Acoustic Basis of Preferences for Infant-Directed Singing," *Infant Behavior and Development* 20 (1997): 383–396.

50. Trehub, "Human Processing Predispositions and Musical Universals."

51. Sandra E. Trehub, "Mothers Are Musical Mentors," *Zero to Three,* September 2002, p. 21.

52. In fact, a recent study shows that six- and seven-month-olds with no previous exposure to test lullabies prefer the lower-pitched rendition. See Volkova, Trehub, and Schellenberg, "Infants' Memory for Musical Performances."

53. Trehub, "Mothers Are Musical Mentors."

54. Ibid.

55. Elena Longhi and Annette Karmiloff-Smith, "In the Beginning Was the Song: The Complex Multimodal Timing of Mother-Infant Musical Interaction," *Behavioral and Brain Sciences* 27 (2004): 516–517.

56. Ibid., p. 517.

57. By the time they are three months old, babies display synchronous behaviors with the stressed more than unstressed segments of their mothers' songs.

58. Dissanayake, "If Music Is the Food of Love"; also see Ellen Dissanayake, "Antecedents of the Temporal Arts in Early Mother-Infant Interaction," in *The Origins of Music,* Wallin, Merker, and Brown, eds.

59. Jessica Phillips-Silver and Laurel J. Trainor, "Feeling the Beat: Movement Influences Infant Rhythm Perception," *Science* 308 (2005): 1430.

60. This makes perfect sense because the vestibular system that is concerned with body orientation and balance includes the semicircular canals that are part of the inner ear.

61. J. B. Morton and Sandra Trehub, "Children's Understanding of Emotion in Speech," *Child Development* 72 (2001): 834–843.

62. For discussion, see ibid.

63. Andrew Whiten and Richard W. Byrne, *Machiavellian Intelligence II: Extensions and Evaluations* (New York: Cambridge University Press, 1997).

64. Would *you* want to have lunch with Hannibal Lecter?

65. Dean Falk, "Hominid Brain Evolution and the Origin of Music," in *The Origins of Music,* Wallin, Merker, and Brown, eds.

66. Trehub, "Mothers Are Musical Mentors"; Trainor, Clark, Huntley, and Adams, "The Acoustic Basis of Preferences for Infant-Directed Singing."

67. For details, see Falk, *Braindance.*

68. F. H. Rauscher and G. L. Shaw, "Key Components of the Mozart Effect," *Perceptual and Motor Skills 86* (1998): 835–841; F. H. Rauscher, "Mozart and the Mind: Factual and Fictional Effects of Musical Enrichment," in *Improving Academic Achievement: Impact of Psychological Factors on Education,* J. Aronson, ed. (New York: Academic Press, 2002); Y. Ho, M. Cheung, and A. S. Chan, "Music Training Improves Verbal But Not Visual Memory: Cross-Sectional and Longitudinal Explorations in Children," *Neuropsychology* 17 (2003): 439–450.

69. Jayne M. Standley, "A Meta-Analysis of the Efficacy of Music Therapy for Premature Infants," *Journal of Pediatric Nursing* 17 (2002): 107–113.

CHAPTER 8: ANCIENT GESTURES, MODERN ART

1. For a classic paper on the gestural hypothesis, see G. W. Hewes, "Primate Communication and the Gestural Origin of Language," *Current Anthropology* 14 (1973): 5–24.

2. A. S. Pollick and Frans B. M. de Waal, "Ape Gestures and Language Evolution," *Proceedings of the National Academy of Sciences USA* 104 (207): 8184–8189.

3. Dean Falk, *Primate Diversity* (New York: W. W. Norton, 2000), p. 203.

4. For discussion, see Dean Falk, *Braindance,* revised and expanded ed. (Gainesville: University Press of Florida, 2004).

5. J. M. Iverson and Susan Goldin-Meadow, "Gesture Paves the Way for Language Development," *Psychological Science* 16 (2005): 367–371.

6. P. Greenfield and J. Smith, *The Structure of Communication in Early Language Development* (New York: Academic Press, 1976).

7. Iverson and Goldin-Meadow, "Gesture Paves the Way for Language Development."

8. Ibid.

9. David F. Armstrong, William C. Stokoe, and Sherman E. Wilcox, "Signs of the Origin of Syntax," *Current Anthropology* 35 (1994): 349–368.

10. Ibid.

11. Ibid., p. 355.

12. Ibid., p. 355.

13. In more formal linguistic terms, this subject, verb, and direct object would be called the agent, action, and patient.

14. Armstrong, Stokoe, and Wilcox, "Signs of the Origin of Syntax."

15. Adam Kendon, "How Gestures Can Become Like Words," in *Cross-cultural Perspectives in Nonverbal Communication,* F. Poyatos, ed. (Toronto: Hogrefe, 1988), pp. 131–141.

16. Armstrong, Stokoe, and Wilcox, "Signs of the Origin of Syntax," p. 350.

17. Armstrong, Stokoe, and Wilcox also include vocal gestures in their four levels, but these are peripheral to the present discussion, which for the moment focuses on nonvocal gestures.

18. For a review, see Chapter 7 of Nobuo Masataka, *The Onset of Language* (Cambridge, UK: Cambridge University Press, 2003).

19. There are some differences, too. In particular, the space in front of signers' bodies (the *sign space*) plays a special role in the grammar of sign languages.

20. Susan Goldin-Meadow and C. Mylander, "Spontaneous Sign Systems Created by Deaf Children in Two Cultures," *Nature* 391 (1998): 279–281.

21. A. Senghas, S. Kita, and A. Ozyurek, "Children Creating Core Properties of Language: Evidence from an Emerging Sign Language in Nicaragua," *Science* 305 (2004): 1779–1782.

22. A similar process occurs in the formation of spoken *pidgins*, which incorporate words from two languages.

23. Senghas, Kita, and Ozyurek, "Children Creating Core Properties of Language," p. 1781.

24. Ibid., p. 1781.

25. Nobuo Masataka, "Perception of Motherese in Japanese Sign Language by 6-Month-Old Hearing Infants," *Developmental Psychology* 34 (1998): 241.

26. Laura Ann Petitto, S. Holowka, L. E. Sergio, and D. Ostry, "Language Rhythms in Baby Hand Movements," *Nature* 413 (2001): 35–36.

27. Senghas, Kita, and Ozyurek, "Children Creating Core Properties of Language," p. 1781.

28. Ibid.

29. Neuroscientists have long recognized divisions of the entire nervous system (including the brain) into sensory and motor systems (although they are

connected and sometimes overlap). In general, sensory systems are located cau-
dal to (in back of) motor systems but, again, with some overlap.

30. M. Myowa-Yamakoshi, M. Tomonaga, M. Tanaka, and T. Matsuzawa, "Imitation in Neonatal Chimpanzees (*Pan Troglodytes*)," *Developmental Science* 7 (2004): 437–442.

31. Pier F. Ferrari, E. Visalberghi, A. Paukner, L. Fogassi, A. Ruggiero, and S. J. Suomi, "Neonatal Imitation in Rhesus Macaques," *PLoS Biology* 4 (2006): e302.

32. Such early imitation of gestures by infant primates is very interesting in light of the startling discovery that macaques have cells in their brains that fire not only when a monkey engages in a specific act (such as reaching for food), but also when the monkey sees another individual perform the same behavior (G. Rizzolatti and L. Craighero, "The Mirror-Neuron System," *Annual Review of Neuroscience* 27 [2004]: 169–192). Because activity in these cells mirrors the behavior of other individuals, they are called *mirror neurons*. And they aren't just associated with vision. Monkeys also have auditory mirror neurons that fire when they produce a particular sound or hear another individual do so—for example, by tearing a piece of paper. Humans have mirror neurons, too, which makes these special cells of even greater interest to evolutionary biologists (more on this in Chapter 9) (Dean Falk, "Prelinguistic Evolution in Early Ho-minins: Whence Motherese?" *Behavioral and Brain Sciences* 27 [2004]: 491–541). Indeed, various workers have suggested that mirror neurons may provide the neurological basis for understanding actions in others and are, in fact, part of an action-perception network that facilitates both facial and manual gestures, as well as (in the case of humans, at least) the emergent linguistic communication between mothers and infants.

33. Andrew N. Meltzoff and J. Decety, "What Imitation Tells Us About So-cial Cognition: A Rapprochement Between Developmental Psychology and Cognitive Neuroscience," *Philosophical Transactions of the Royal Society of London B* 358 (2003): 491.

34. "For example, there is an intimate relation between 'striving to achieve a goal' and the concomitant facial expression and effortful bodily acts. Infants experience their own unfulfilled desires and their own concomitant facial/postural/vocal reactions. They experience their own inner feelings and outward facial expressions and construct a detailed bidirectional map linking mental ex-periences and behaviour" (Meltzoff and Decety, "What Imitation Tells Us About Social Cognition," p. 491.

35. Ibid., p. 497.

36. Meltzoff hypothesizes that "infants can represent human movement patterns they see and ones they perform using the same mental code. There is thus something like an act space or primitive body scheme that allows the in-fant to unify the visual and motor/proprioceptive information into one com-mon 'supramodal' framework. This supramodal act space is not restricted to modality-specific information (visual, tactile, motor, etc.)." (Andrew Meltzoff, "'Like Me': A Foundation for Social Cognition," *Developmental Science* 10 [2007]: 130.)

37. Merlin Donald, "Preconditions for the Evolution of Protolanguages," in *The Descent of Mind: Psychological Perspectives on Hominid Evolution,* M. C. Corballis and I. Lea, eds. (Oxford, UK: Oxford University Press, 1999), pp. 138–151.

38. Ibid.

39. Donald identifies the process for generating motor action-patterns as one that "relies on a principle of perceptual resemblance," which he accordingly labels "mimesis" or "mimetic skill." He also claims that mimesis is representational without being linguistic and that it underlies mime, body language, gesture, most other nonverbal communication, ritual, some music, and dance. A compelling manifestation of mimesis is found in motor rhythms, which may be transferred to various parts of the body. Thus, an auditory, proprioceptive, or visual source of rhythm can be tracked with the voice, feet, whole body, or fingers. "For instance, in a jazz drummer's improvisations; in dance or marching; in coordinated group song; in many children's games; or in gymnastics." Mimesis alone was not enough to produce protolanguage, of course. (Donald, "Preconditions for the Evolution of Protolanguages," p. 145.)

40. Ibid., p. 142.

41. Armstrong, Stokoe, and Wilcox, "Signs of the Origin of Syntax."

42. Sherman Wilcox, "Language from Gesture," *Behavioral and Brain Sciences* 27 (2004): 526.

43. J. M. Iverson, O. Capirci, E. Longobardi, and M. C. Caselli, "Gesturing in Mother-Child Interactions," *Cognitive Development* 14 (1999): 57–75.

44. L. J. Gogate, L. E. Bahrick, and J. D. Watson, "A Study of Multimodal Motherese: The Role of Temporal Synchrony Between Verbal Labels and Gestures," *Child Development* 71 (2000): 878–894.

45. K. L. Schmidt and J. F. Cohn, "Human Facial Expressions as Adaptations: Evolutionary Questions in Facial Expression Research," *Yearbook of Physical Anthropology* 44 (2001): 3–24.

46. T. Rome-Flanders and C. Cronk, "A Longitudinal Study of Infant Vocalizations During Mother-Infant Games," *Journal of Child Language* 22 (1995): 259–274.

47. M. Tomasello and M. Carpenter, "Shared Intentionality," *Developmental Science* 10 (2007): 121–125; E. S. Messinger and A. Fogel, "Give and Take: The Development of Conventional Infant Gestures," *Merrill-Palmer Quarterly* 44 (1998): 566–590.

48. C. F. Papaeliou and C. Trevarthen, "Prelinguistic Pitch Patterns Expressing 'Communication' and 'Apprehension,'" *Journal of Child Language* 33 (2006): 163–178.

49. For details, see Falk, "Prelinguistic Evolution in Early Hominins," pp. 497–498.

50. As always, the best models for our earliest ancestors are chimpanzees, and their gestures differ from those of humans in several important ways (M. Tomasello and L. Camaioni, "A Comparison of the Gestural Communication of Apes and Human Infants," *Human Development* 40 [1997]: 7–24). Chimpanzees use gestures almost exclusively to attract attention to themselves or to get another individual to do what they want. Their gestures also usually involve physical contact between the gesturer and the recipient. Such gestures are said to be

proximal rather than distal. As human infants begin to mature, on the other hand, their gestures often refer to objects or external events rather than to themselves, and they usually do not physically contact the individual they are addressing.

51. R. Arnheim, *Art and Visual Perception: A Psychology of the Creative Eye* (London: Faber & Faber, 1956).

52. Glyn V. Thomas and Angèl M. J. Silk, *An Introduction to the Psychology of Children's Drawings* (New York: Harvester Wheatsheaf, 1990).

53. M. Tanaka, M. Tomonaga, and T. Matsuzawa, "Finger Drawing by Infant Chimpanzees (*Pan Troglodytes*)," *Animal Cognition* 6 (2003): 245–251.

54. Unless otherwise stated, this discussion of children's artistic development is based on Thomas and Silk, *An Introduction to the Psychology of Children's Drawings*.

55. Ibid., p. 37 for discussion of tadpole figures.

56. Ibid., pp. 39–40, 66–68 for discussions of universal elements in children's drawings.

57. R. Arnheim, *Visual Thinking* (London: Faber & Faber, 1969).

58. According to Thomas and Silk, "We should not underestimate the difficulties that the construction of such combinations can present. Such constructed drawings have to be planned: first, decisions have to be taken about the order of drawing the component parts; second, when positioning the first-drawn elements, space has to be reserved for parts which will be added later in the drawing sequence; finally, later-drawn elements of the picture have to be positioned so that they can be joined on to previously drawn elements as appropriate," p. 77.

59. Kyoko Yamagata, "Emergence of representational activity during the early drawing stage: process analysis," *Japanese Psychological Research* 43 (2001): 130–140.

60. Ibid., p. 138.

61. Susan Rich Sheridan, "A Theory of Marks and Mind: The Effect of Notational Systems on Hominid Brain Evolution and Child Development with an Emphasis on Exchanges Between Mothers and Children," *Medical Hypotheses* 64 (2005): 417–427.

62. Ibid., p. 423.

63. An excellent resource for viewing prehistoric art is www.origins net.org. James B. Harrod is the manager and webmaster

64. Robert G. Bednarik, "The 'Australopithecine' Cobble from Makapansgat, South Africa," *South African Archaeological Bulletin* 53 (1998): 4–8.

65. Robert G. Bednarik, G. Kumar, A. Watchman, and R. G. Roberts, "Preliminary Results of the EIP [Early Indian Petroglyphs] Project," *Rock Art Research* 22, 2 (2005): 147–197.

66. Thomas and Silk, *An Introduction to the Psychology of Children's Drawings.*

67. D. Mania and U. Mania, "Deliberate Engravings on Bone Artefacts of *Homo erectus*," *Rock Art Research* 5,2 (1988): 91–95.

68. Robert G. Bednarik, "Concept-Mediated Marking in the Lower Palaeolithic," *Current Anthropology* 36 (1995): 605–634.

69. Ibid., p. 613.

70. Ibid., p. 614.

71. Robert G. Bednarik, "Neurophysiology and Paleoart," Lecture No. 6, Semiotix Course 2006, Cognition and Symbolism in Human Evolution, www.chass.utoronto.ca/epc/srb/cyber/rbednarik6.pdf.

72. R. Kellogg, M. Knoll, and J. Kugler, "Form-Similarity Between Phosphenes of Adults and Pre-School Children's Scribblings," *Nature* 208 (1965): 1129–1130.

73. Bednarik, "Neurophysiology and Paleoart," p. 7.

74. Robert G. Bednarik, "The Nature of Psychograms," *The Artefact* 8 (1984): 27–32.

75. By 200,000 years ago, for example, hominins in Germany were producing little sculptures of mammoths, elephants, wild boars, rhinoceroses, sitting birds, flying birds, and fish, in addition to hominin faces and masks. (Many illustrations of early hominin portrayals of animals are provided at www.origins net.org.)

76. Bednarik, who has studied the image, observes: "As a scientist I have no desire to speculate about its meaning or purpose, the creation of archaeological myths is the domain of archaeologists. Naturally some will see a complex vulva design (a traditional favourite), others an arrow, or bird tracks, and others again will see a stickman. There may well be merit in one of these speculations (and any of the others I can think of). In particular there can be no doubt that, were this motif occurring in rock art, it would certainly be described as a human figure, and indeed as a male human figure (the slight extension of the central line beyond where it meets two others would then be seen as a penis)." (Robert G. Bednarik, "The Middle Paleolithic Engravings from Oldisleben, Germany," *Anthropologie* 44, no. 2 [2006]: 118.)

77. For example, half-human, half-lion figures ("lion men") appear at this time, including one from a cave in southwestern Germany that is more than thirty thousand years old (N. J. Conrad, "Palaeolithic ivory sculptures from southwestern Germany and the origins of figurative art," *Nature* 246 [2003]: 830–832). The walls of the famous Grotte Chauvet in southern France are from another important site, which contains a colorful menagerie that includes reindeer, mammoths, horses, bison, bears, lions, rhinoceros, a red panther, and an engraved owl. These images are so exquisite and skillfully executed (some of them in bas-relief) that they look as if they were created by highly skilled modern artists.

78. For discussion of Venus figurines and other art, see Falk, *Braindance*.

79. For example, recent evidence suggests that humans who lived at least 100,000 years ago in Israel and North Africa were wearing beads made of pierced shells that had been transported from distant seashores. (M. Vanhaeren, F. d'Errico, C. Stringer, S. L. James, J. A. Todd, and H. K. Mienis, "Middle Paleolithic Shell Beads in Israel and Algeria," *Science* 312 [2006]: 1785–1788.) This is highly significant because use of such personal adornments is generally interpreted as an indication of symbolic communication.

80. K. Sharpe and L. Van Gelder, "Evidence for Cave Markings by Palaeolithic Children," *Antiquity* 80 (2006): 937–947; R. D. Guthrie, *The Nature of Paleolithic Art* (Chicago: University of Chicago Press, 2006).

81. J. Berck, "Before Baby Talk, Signs and Signals," *New York Times,* January 6, 2004.

82. S. R. Sheridan, "Very Young Children's Drawings and Human Consciousness: The Scribble Hypothesis. A Plea for Brain-Compatible Teaching and Learning," poster presentation, Toward a Science of Consciousness Conference, Skovde, Sweden, August 2001.

CHAPTER 9: FINDING OUR TONGUES

1. This is Steven Pinker's famous expression. Steven Pinker, *How the Mind Works* (New York: W. W. Norton, 1997), p. 534.

2. Some think most of the increase in brain size before 600,000 years ago had to do with enlarging bodies (C. B. Ruff, E. Trinkaus, and T. W. Holliday, "Body Mass and Encephalization in Pleistocene *Homo,*" *Nature* 387 [1997]: 173–176), while others believe hominin brains simply got bigger in fits and starts.

3. For discussion and references, see Dean Falk, "Evolution of the Primate Brain," in *Handbook of Paleoanthropology, Volume II Primate Evolution and Human Origins,* W. Henke and I. Tattersall, eds. (Berlin: Springer-Verlag, 2007).

4. It appears, however, that brains and bodies did not enlarge at exactly the same rate during hominin evolution because, as detailed in Chapter 4, a nearly complete skeleton of a *Homo erectus* boy (WT 15000) who lived in Kenya some 1.6 million years ago indicates that he had a modern body size, while his projected adult cranial capacity of around 900 cm^3 is only two-thirds the average volume for modern folks. This suggests that the increase in body size may have been a bit ahead of that for brains.

5. M. A. Hofman, "Brain Evolution in Hominids: Are We at the End of the Road?" in *Evolutionary Anatomy of the Primate Cerebral Cortex,* D. Falk and K. R. Gibson, eds. (Cambridge, UK: Cambridge University Press, 2001).

6. Despite the fact that they had the largest average cranial capacity on record, many workers believe that Neanderthals were not in the lineage that led to *Homo sapiens*.

7. R. A. Dart, "The Relationship of Brain Size and Brain Pattern to Human Status," *South African Journal of Medical Science* 21 (1956): 23–45.

8. See, for example, Dean Falk, *Braindance,* revised and expanded ed. (Gainesville: University Press of Florida, 2004). For a current overview of primate brain evolution from the scientific literature, see Falk, "Evolution of the Primate Brain."

9. Relative size refers here to the size of lobes of the brain compared with that of the whole brain.

10. K. Semendeferi, "Advances in the Study of Hominoid Brain Evolution: Magnetic Resonance Imaging (MRI) and 3-D Reconstruction," in *Evolutionary Anatomy of the Primate Cerebral Cortex*.

11. N. F. Dronkers, "A New Brain Region for Coordinating Speech Articulation," *Nature* 384 (1996): 159–161.

12. Todd M. Preuss, "The Discovery of Cerebral Diversity: An Unwelcome Scientific Revolution," in *Evolutionary Anatomy of the Primate Cerebral Cortex*.

See also T. M. Preuss, "Who's Afraid of *Homo sapiens?*" *Journal of Biomedical Discovery and Collaboration* (2006), www.j-biomed-discovery.com/content/1/1/17.

13. T. M. Preuss, "Evolutionary Specializations of Primate Brain Systems," in *Primate Origins: Adaptations and Evolution,* M. J. Ravosa and M. Dagasto, eds. (New York: Springer, 2007).

14. Ibid., p. 662.

15. Ibid., p. 664.

16. I have taken some liberty with this figure—for instance, by putting the tongue in, instead of entirely below, the mouth. The legs and feet would normally extend over the middle of the brain and not be visible from the side.

17. Kuniyoshi Sakai, "Language Acquisition and Brain Development," *Science* 310 (2005): 815–823.

18. S. T. Grafton, L. Fadiga, M. A. Arbib, and Giacomo Rizzolatti, "Premotor Cortex Activation During Observation and Naming of Familiar Tools," *Neuro Image* 6 (1997): 231–236.

19. Ibid.

20. L. L. Chao and A. Martin, "Representation of Manipulable Man-Made Objects in the Dorsal Stream," *NeuroImage* 12 (2000): 478–484; Grafton, Fadiga, Arbib, and Rizzolatti, "Premotor Cortex Activation."

21. Preuss, "Evolutionary Specializations of Primate Brain Systems," p. 664.

22. Sakai, "Language Acquisition and Brain Development."

23. Giacomo Rizzolatti and A. Arbib, "Language Within Our Grasp," *Trends in Neurosciences* 21 (1998): 188–194. Interestingly, these areas often involve voluntary eye movements (E. C. Crosby, T. Humphrey, and E. W. Lauer, *Correlative Anatomy of the Nervous System* [New York: Macmillan, 1962]).

24. Rizzolatti and Arbib, "Language Within Our Grasp."

25. Sound familiar? It should, because this discussion is reminiscent of Meltzoff's interpretation regarding the proclivity of newborn humans to imitate facial gestures observed in adults (Chapter 8).

26. M. Iacoboni, R. P. Woods, M. Brass, H. Bekkering, J. C. Mazziotta, and Giacomo Rizzolatti, "Cortical Mechanisms of Human Imitation," *Science* 286 (1999): 2526–2528.

27. Monkeys and humans also have mirror neurons in their temporal and parietal lobes that respond to seeing and carrying out activities with hands.

28. V. Gazzola, L. Aziz-Zadeh, and C. Keysers, "Empathy and the Somatotopic Auditory Mirror System in Humans," *Current Biology* 16 (2006): 1824–1829.

29. S. M. Wilson, A. S. Saygin, M. I. Sereno, and M. Iacoboni, "Listening to Speech Activates Motor Areas Involved in Speech Production," *Nature Neuroscience* 7 (2004): 701–702.

30. Stein Braten, "Hominin Infant Decentration Hypothesis: Mirror Neurons System Adapted to Subserve Mother-Centered Participation," *Behavioral and Brain Sciences* 27 (2004): 508.

31. Braten calls this "altercentric mirroring and self-with-other resonance."

32. Read, for example, my interview with dancer Cynthia Riffle Bowers in Falk, *Braindance,* pp. 73–78.

33. C. Keysers, B. Wicker, V. Gazzola, J. L. Anton, L. Fogassi, and V. Gallese, "A Touching Sight: SII/PV Activation During the Observation and Experience of Touch," *Neuron* 42 (2004): 335–346.

34. D. L. Cheney and R. M. Seyfarth, *Baboon Metaphysics* (Chicago: University of Chicago Press, 2007).

35. E. Herrmann, J. Call, M. V. Hernandez-Lloreda, B. Hare, and M. Tomasello, "Humans Have Evolved Specialized Skills of Social Cognition: The Cultural Intelligence Hypothesis," *Science* 317 (2007): 1360–1366.

36. Herrmann, Call, Hernandez-Lloreda, Hare, and Tomasello, "Humans Have Evolved Specialized Skills of Social Cognition," p. 1365.

37. J. K. Hamlin, K. Wynn, and P. Bloom, "Social Evaluation by Preverbal Infants," *Nature* 450 (2007): 557–559.

38. M. Hopkins, "Babies Can Spot Nice and Nasty Characters," *Nature* 450 (2007): www.nature.com/news/2007/071121/full/news.2007.278.html.

39. R. W. Byrne and A. Whiten, eds., *Machiavellian Intelligence: Social Expertise and the Evolution of Intellect in Monkeys, Apes and Humans* (Oxford, UK: Clarendon Press, 1988).

40. R.I.M. Dunbar, "Coevolution of Neocortical Size, Group Size and Language in Humans," *Behavioral and Brain Sciences* 16 (1993): 681–735.

41. As detailed in this book, the "putting the baby down" hypothesis also entails that voices began substituting for hands. Although our hypotheses are consistent with each other, mine focuses on the early emergence of protolanguage, while Dunbar's addresses the more recent emergence of language.

42. Sakai, "Language Acquisition and Brain Development." The graph has been provided courtesy of the author.

43. Gazzola, Aziz-Zadeh, and Keysers, "Empathy and the Somatotopic Auditory Mirror System in Humans." See also L. Fogassi, P. F. Ferrari, B. Gesierich, S. Rozzi, F. Chersi, and Giacomo Rizzolatti, "Parietal Lobe: from Action Organization to Intention Understanding," *Science* 308 (2005): 662–667. It has also been suggested that people with autism (including Asperger's syndrome) may have less than fully functioning mirror neuron systems.

44. N. Mekel-Bobrov and B. T. Lahn, "What Makes Us Human: Revisiting an Age-Old Question in the Genomic Era," *Journal of Biomedical Discovery and Collaboration* 1, no. 18 (2006), doi:10.1186/1747-5333-1-18.

45. E. S. Lander, et al., "Initial Sequencing and Analysis of the Human Genome," *Nature* 409 (2001): 860–921.

46. Chimpanzee Sequencing and Analysis Consortium, "Initial Sequence of the Chimpanzee Genome and Comparison with the Human Genome," *Nature* 437 (2005): 69–87.

47. Mekel-Bobrov and Lahn, "What Makes Us Human."

48. Geneticists have also developed powerful analytical tools to help identify specific genes that may have been a focus of natural selection in our ancestors. For example, forty-seven genes expressed in the human brain have been recognized as likely products of adaptive evolution, although it is not yet clear what exactly these genes do (X. J. Yu, et al., "Detecting Lineage-Specific Adaptive

Evolution of Brain-Expressed Genes in Human Using Rhesus Macaque as Outgroup," *Genomics* 88 [2006]: 745–751.)

49. Microcephaly, for example, is a pathology in which people are born with small and somewhat misshapen brains and heads, and it is usually accompanied by mild to severe mental impairment. It is also known to be a "grab bag" disease that can result from a variety of environmental or genetic causes. Some geneticists have argued that two of the genes (*ASPM* and *Microcephalin*) that are sometimes associated with microcephaly may point to parts of the human genome that contributed to the evolution of brain size, i.e., when the relevant genes appeared in normal rather than mutated disease-causing forms (Mekel-Bobrov and Lahn, "What Makes Us Human"). This is an intriguing idea, although it is important to keep in mind that hundreds of other genes have also been associated with neurological development, so it would be too simplistic to say that these two microcephalic genes point to parts of the genome that were exclusively involved in brain size evolution.

50. F. Liegeois, et al., "Language fMRI Abnormalities Associated with *FOXP2* Gene Mutation," *Nature Neuroscience* 6 (2003): 1230–1237.

51. Interestingly, mice that have disrupted *FOXP2* genes fail to make the ultrasonic sounds that young mice produce when they are separated from their mothers, perhaps because of deficits in fine motor coordination, according to Joseph Buxbaum of the Mount Sinai School (W. Shu et al., "Altered Ultrasonic Vocalization in Mice with a Disruption in the FOXP2 Gene." *Proceedings of the National Academy of Sciences, USA* 102 (2005): 9643–9648.

52. M. C. Corbalis, "*FOXP2* and the Mirror System," *Trends in Cognitive Sciences* 8 (2004): 96.

BIBLIOGRAPHY

Adovasio, J. M., Soffer, O., and Page, J. 2007. *The Invisible Sex: Uncovering the True Roles of Women in Prehistory.* New York: Harper Collins.

Alemseged, Z., Spoor, F., Kimbel, W. H., Bobe, R., Geraads, D., Reed, D., and Wynn, J. G. 2006. "A Juvenile Early Hominin Skeleton from Dikika, Ethiopia." *Nature* 443(7109):296–301.

Andruski, J., Kuhl, P. K., and Hayashi, A. 1999. "Point Vowels in Japanese Mothers' Speech to Infants and Adults." *The Journal of the Acoustical Society of America* 105(2):1095–1096.

Armstrong, D. F., Stokoe, W. C., and Wilcox, S. E. 1994. "Signs of the Origin of Syntax." *Current Anthropology* 35(4):349–368.

Arnheim, R. 1956. *Art and Visual Perception: A Psychology of the Creative Eye.* London: Faber and Faber.

Arnheim, R. 1969. *Visual Thinking.* Berkeley: University of California Press.

Arthur, W. 2007. "The Search for Novelty." *Nature* 447(7142):261–262.

Balter, M. 2007. "Brain, Music, and Sound Research Center. Study of Music and the Mind Hits a High Note in Montreal." *Science* 315(5813):758–759.

Balzamo, E., Bradley, R. J., and Rhodes, J. M. 1972. "Sleep Ontogeny in the Chimpanzee: From Two Months to Forty-One Months." *Electroencephalography and Clinical Neurophysiology* 33(1):47–60.

Bard, K. A. 2004. "What Is the Evolutionary Basis for Colic?" *Behavioral and Brain Sciences* 27:459.

Bavin, E. 1995. "Language Acquisition in Crosslinguistic Perspective." *Annual Review of Anthropology* 24:373–396.

Bednarik, R. G. 1984. "On the Nature of Psychograms." *The Artefact* 8:27–33.

Bednarik, R. G. 1998. "The 'Australopithecine' Cobble from Makapansgat." *South African Archaeological Bulletin* 53:4–8.

Bednarik, R. G. 1995. "Concept-Mediated Marking in the Lower Palaeolithic." *Current Anthropology* 36(4):605–634.

Bednarik, R. G. 2006. Neurophysiology and Paleoart, Lecture No. 6, Semiotix Course 2006, Cogniton and Symbolism in Human Evolution. www.chass.utoronto.ca/epc/srb/cyber/rbednarik6.pdf.

Bednarik, R. G. 2006. "The Middle Palaeolithic Engravings from Oldisleben, Germany." *Anthropologie* 44(2):113–121.

Bednarik, R. G., Kumar, G., Watchman, A., and Roberts, R. G. 2005. "Preliminary Results of the EIP (Early Indian Petroglyphs) Project." *Rock Art Research* 22(2):147–197.

Bell, S. M., and Ainsworth, M. D. 1972. "Infant Crying and Maternal Responsiveness." *Child Development* 43(4):1171–1190.

Berck, J. 2004, Jan. 6. "Before Baby Talk, Signs, and Signals." *New York Times.*

Biben, M., Symmes, D., and Bernhards, D. 1989. "Contour Variables in Vocal Communication Between Squirrel Monkey Mothers and Infants." *Developmental Psychobiology* 22(6):617–631.

Bickerton, D. 2004. "Vocalizations Don't Equal Language." *Behavioral and Brain Sciences* 27:504–505.

Bickerton, D. 2007. "Language Evolution: A Brief Guide for Linguists." *Lingua* 117(3):510–526.

Bloom, P. 1999. "Language Capacities: Is Grammar Special?" *Current Biology* 9(4):R127–128.

Bloom, P. 2000. *How Children Learn the Meaning of Words.* Cambridge, MA: MIT Press.

Blurton-Jones, N. G. 1993. "The Lives of Hunter-Gatherer Children: Effects of Parental Behavior and Parental Reproductive Strategy." In M. Pereira M and L. A. Fairbanks, eds., *Juvenile Primates.* Oxford: Oxford University Press.

Boesch, C., and Boesch-Achermann, H. 1991, Sept. "Dim Forest, Bright Chimps." *Natural History*:50–57.

Bowlby, J. 1982. *Attachment and Loss: Vol. 1.* New York: Basic Books.

Bowlby, J. 1982. "Attachment and Loss: Retrospect and Prospect." *American Journal of Orthopsychiatry* 52(4):664–678.

Braine, M. D. 1976. "Children's First Word Combinations." Monographs of the Society for Research in Child Development 41.

Bramble, D. M., and Lieberman, D. E. 2004. "Endurance Running and the Evolution of Homo." *Nature* 432(7015):345–352.

Braten, S. 2004. "Hominin Infant Decentration Hypothesis: Mirror Neurons System Adapted to Subserve Mother-Centered Participation." *Behavioral and Brain Sciences* 27:508–509.

Brown, P., Sutikna, T., Morwood, M. J., Soejono, R. P., Jatmiko, Saptomo E. W., and Due, R. A. 2004. "A New Small-Bodied Hominin from the Late Pleistocene of Flores, Indonesia." *Nature* 431(7012):1055–1061.

Brown, R. 1973. *A First Language: The Early Stages.* Cambridge, MA: Harvard University Press.

Burling, R. 1993. "Primate Calls, Human Language, and Nonverbal Communication." *Current Anthropology* 34(1):25–37.

Burling, R. 2004. "Prosody Does Not Equal Language." *Behavioral and Brain Sciences* 27:509.

Burnham, D., Kitamura, C., and Vollmer-Conna, U. 2002. "What's New, Pussycat? On Talking to Babies and Animals." *Science* 296(5572):1435.

Byrne, R. W., and Whiten, A. 1988. *Machiavellian Intelligence: Social Expertise and the Evolution of Intellect in Monkeys, Apes, and Humans.* New York and Oxford: Clarendon Press/Oxford University Press.

Cameron-Faulkner, T., Lieven, E., and Tomasello, M. 2003. "A Construction-Based Analysis of Child-Directed Speech." *Cognitive Science 27*:843–873.

Carey, B. 2007, Oct. 23. "An Active, Purposeful Machine That Comes Out at Night to Play." *New York Times.*

Case, C. 2005. "All That's Glitter and Golden: The Story of Gombe's Famous Twins." Jane Goodall Institute News Center. www.janegoodall.org/news.

Chao, L. L., and Martin, A. 2000. "Representation of Manipulable Man-Made Objects in the Dorsal Stream." *Neuroimage 12*(4):478–484.

Cheney, D. L., and Seyfarth, R. M. 2007. *Baboon Metaphysics: The Evolution of a Social Mind.* Chicago: University of Chicago Press.

Cheney, D. L., and Seyfarth, R. M. 1990. *How Monkeys See The World: Inside the Mind of Another Species.* Chicago: University of Chicago Press.

Chomsky, N. 1986. *Knowledge of Language: Its Nature, Origin, and Use.* New York: Praeger.

Christensson, K., Cabrera, T., Christensson, E., Uvnas-Moberg, K., and Winberg, J. 1995. "Separation Distress Call in the Human Neonate in the Absence of Maternal Body Contact." *Acta Paediatrica 84*(5):468–473.

Conrad, N. 2003. "Paleolithic Ivory Sculptures from South-Western Germany and the Origins of Figurative Art." *Nature 426*:830–832.

Consortium CSaA. 2005. "Initial Sequence of the Chimpanzee Genome and Comparison with the Human Genome." *Nature 437*:69–87.

Cooke, D. 1959. *The Language of Music.* London and New York: Oxford University Press.

Coolidge, F., and Wynn, T. 2006. "The Effects of the Tree-to-Ground Transition in the Evolution of Cognition in Early Homo." *Before Farming 4,* article 11.

Cooper, R., Abraham, J., Berman, S., and Staska, M. 1997. "The Development of Infants' Preference for Motherese." *Infant Behavior and Development 20*(4):477–488.

Corballis, M. C. 2004. "FOXP2 and the Mirror System." *Trends in Cognitive Sciences 8*(3):95–96.

Cordes, I. 2003. "Melodic Contours as a Connecting Link Between Primate Communication and Human Singing." In R. Kopiez, A. C. Lehmann, I. Wolther, and C. Wolf, eds., *The Proceedings of the 5th Triennial European Society for the Cognitive Sciences of Music (ESCOM) Conference.* Hanover, Germany: Hanover University of Music and Drama.

Cowley, S., Moodley, S., and Fiori-Cowley, A. 2004. "Grounding Signs of Culture: Primary Intersubjectivity in Social Semiosis." *Mind Culture and Activity 11*(2):109–132.

Crittenden, A. N, and Marlowe, F. 2008. "Allomaternal Care Among the Hadza of Tanzania." *Human Nature: An Interdisciplinary Biosocial Perspective 19*(3):249–262.

Crockford, C., Herbinger, I., Vigilant, L., and Boesch, C. 2004. "Wild Chimpanzees Produce Group-Specific Calls: A Case for Vocal Learning?" *Ethology 110*(3):221–243.

Crosby, E., Humphrey, T., and Lauer, E. W. 1962. *Correlative Anatomy of the Nervous System.* New York: Macmillan.

Dart, R. A. 1956. "The Relationships of Brain size and Brain Pattern to Human Status." *South African Journal of Medical Sciences* 21(1–2):23–45.

Darwin, C. 1859. *On the Origin of Species by Means of Natural Selection.* London: J. Murray.

Darwin, C. 1871. *The Descent of Man, and Selection in Relation to Sex.* New York: D. Appleton.

deBoer, B., and Kuhl, P. K. 2003. "Investigating the Role of Infant-Directed Speech with a Computer Model." *Acoustics Research Letters Online:* 129–134.

Delaney C. 2000. "Making Babies in a Turkish Village." In J. S. DeLoache and A. Gottlieb, eds., *A World of Babies: Imagined Childcare Guides for Seven Societies.* Cambridge: Cambridge University Press.

DeLoache, J. S, and Gottlieb, A. 2000. *A World of Babies: Imagined Childcare Guides for Seven Societies.* Cambridge and New York: Cambridge University Press.

deMenocal, P. 2004. "African Climate Change and Faunal Evolution During the Pliocene-Pleistocene." *Earth and Planetary Science Letters* 220:3–24.

DeSilva, J. M, and Lesnik, J. J. In press. "Brain Size at Birth Throughout Human Evolution: A New Method for Estimating Neonatal Brain Size in Hominins." *Journal of Human Evolution.*

de Waal, F. 2002, Dec. 6. "Before Jane Goodall, There Was Nadia Kohts." *Chronicle of Higher Education.*

de Waal, F.B.M. 1996. *Good Natured: The Origins of Right and Wrong in Humans and Other Animals.* Cambridge, MA: Harvard University Press.

Diener, M. 2000. "Gift from the Gods, a Balinese Guide to Early Child Rearing." In J. S. DeLoache and A. Gottlieb, eds., *A World of Babies: Imagined Childcare Guides for Seven Societies.* Cambridge and New York: Cambridge University Press.

Dissanayake, E. 2000. "Antecedents of the Temporal Arts in Early Mother-Infant Interaction. In N. Wallin, B. Merker, and S. Brown, eds., *The Origins of Music.* Cambridge, MA: MIT Press.

Dissanayake, E. 2008. "If Music Is the Food of Love, What About Survival and Reproductive Success?" *Musicae Scientiae* (special issue 2008):169–195.

Donald, M. 1999. "Preconditions for the Evoluton of Protolanguages." In M. Corballis and I. Lea, eds., *The Descent of Mind: Psychological Perspectives on Hominid Evolution.* Oxford, UK: Oxford University Press.

Draper, P. 1976. "Social and Economic Constraints on Child Life Among the !Kung." In R. Lee and I. DeVore, eds., *Kalahari Hunter-Gatherers.* Cambridge, MA: Harvard University Press.

Dronkers, N. F. 1996. "A New Brain Region for Coordinating Speech Articulation." *Nature* 384(6605):159–161.

Dunbar, R. 1993. "Coevolution of Neocortical Size, Group Size, and Language in Humans." *Behavioral and Brain Sciences* 16:681–735.

Eibl-Eibesfeldt, I. 1989. *Human Ethology.* New York: Aldine de Gruyter.

Everett, D. 2005. "Cultural Constraints on Grammar and Cognition in Piraha." *Current Anthropology* 46:621–646.

Falk, D. 2000. "Hominid Brain Evolution and the Origin of Music." In N. Wallin, B. Merker, and S. Brown, eds., *The Origins of Music*. Cambridge, MA: MIT Press.

Falk, D. 2004. *Braindance* (rev. ed.). Gainesville: University of Florida Press.

Falk, D. 2004. "Prelinguistic Evolution in Early Hominins: Whence Motherese?" *Behavioral and Brain Sciences* 27(4):491–503.

Falk, D. 2007. "Evolution of the Primate Brain. In W. Henke, H. Rothe, and I. Tattersall, eds., *Handbook of Palaeoanthropology: Vol. 2. Primate Evolution and Human Origins*. Berlin and Heidelberg: Springer-Verlag.

Falk, D. 2000. Primate Diversity. New York: W. W. Norton.

Falk, D, Hildebolt, C, Smith, K., Morwood, M. J., Sutikna, T., Brown, P., Jatmiko, Saptomo, E. W., Brunsden, B., and Prior, F. 2005. "The Brain of LB1, Homo Floresiensis." *Science 308*(5719):242–245.

Farrar, M. J. 1990. "Discourse and the Acquisition of Grammatical Morphemes." *Journal of Child Language* 17(3):607–624.

Ferguson, C. A. 1978. "Talking to Children: A Search for Universals." In J. Greenberg, ed., *Universals of Human Language*. Stanford, CA: Stanford University Press.

Fernald, A. 1994. "Human Maternal Vocalizations to Infants as Biologically Relevant Signals: An Evolutionary Perspective." In P. Bloom, ed., *Language Acquisition Core Readings*. Cambridge, MA: MIT Press.

Fernald, A., and Hurtado, N. 2006. "Names in Frames: Infants Interpret Words in Sentence Frames Faster Than Words in Isolation." *Developmental Science* 9(3):F33–40.

Fernald, A., and Morikawa, H. 1993. "Common Themes and Cultural Variations in Japanese and American Mothers' Speech to Infants." *Child Development 64*(3):637–656.

Ferrari, P. F., Visalberghi, E., Paukner, A., Fogassi, L., Ruggiero, A., and Suomi, S. J. 2006. "Neonatal Imitation in Rhesus Macaques." *PLoS Biology 4*(9):e302.

Finlay, B. L., and Darlington, R. B. 1995. "Linked Regularities in the Development and Evolution of Mammalian Brains." *Science 268*(5217):1578–1584.

Fitch, W. T. 2005. "The Evolution of Music in Comparative Perspective." *Annals of the New York Academy of Sciences 1060*:29–49.

Fitch, W. T., Hauser, M. D., and Chomsky, N. 2005. "The Evolution of the Language Faculty: Clarifications and Implications." *Cognition 97*(2):179–210.

Fogassi, L., Ferrari, P. F., Gesierich, B., Rozzi, S., Chersi, F., and Rizzolatti, G. 2005. "Parietal Lobe: From Action Organization to Intention Understanding." *Science 308*(5722):662–667.

Fossey, D. 1983. *Gorillas in the Mist*. Boston, MA: Houghton Mifflin.

Franklin, M. S., and Zyphur, M. J. 2005. "The Role of Dreams in the Evolution of Mind." *Evolutionary Psychology 3*:59–78.

Frodi, A. 1985. "When Empathy Fails." In B. Lester and C.F.Z. Boukydis, eds., *Infant Crying: Theoretical and Research Perspectives*. New York: Plenum.

Ganger, J., and Brent, M. R. 2004. "Reexamining the Vocabulary Spurt." *Developmental Psychology 40*(4):621–632.

Gazzola, V., Aziz-Zadeh, L., and Keysers, C. 2006. "Empathy and the Somatotopic Auditory Mirror System in Humans." *Current Biology 16*(18):1824–1829.

Geissmann, T. 2000. "Gibbon Songs and Human Music from an Evolutionary Perspective." In N. Wallin, B. Merker, and S. Brown, eds., *The Origins of Music.* Cambridge, MA: MIT Press.

Gentner, T. Q., Fenn, K. M., Margoliash, D., and Nusbaum, H. C. 2006. "Recursive Syntactic Pattern Learning by Songbirds." *Nature 440*(7088):1204–1207.

Gogate, L. J., Bahrick, L. E., and Watson, J. D. 2000. "A Study of Multimodal Motherese: The Role of Temporal Synchrony Between Verbal Labels and Gestures." *Child Development 71*(4):878–894.

Goldfield, B. A., and Reznick, J. S. 1990. "Early Lexical Acquisition: Rate, Content, and the Vocabulary Spurt." *Journal of Child Language 17*(1):171–183.

Goldin-Meadow, S., and Mylander, C. 1998. "Spontaneous Sign Systems Created by Deaf Children in Two Cultures." *Nature 391*(6664):279–281.

Goldman, H. I. 2001. "Parental Reports of 'MAMA' Sounds in Infants: An Exploratory Study." *Journal of Child Language 28*(2):497–506.

Golinkoff, R. M., Hirsh-Pasek, K., Cauley, K. M., and Gordon, L. 1987. "The Eyes Have It: Lexical and Syntactic Comprehension in a New Paradigm." *Journal of Child Language 14*(1):23–45.

Goodall J. 1986. *The Chimpanzees of Gombe: Patterns of Behavior.* Cambridge, MA: Belknap Press of Harvard University Press.

Goodall, J. 1990. *Through a Window: My Thirty Years with the Chimpanzees of Gombe.* Boston: Houghton Mifflin.

Goodman, C. S., and Coughlin, B. C. 2000. "Introduction. The Evolution of Evo-Devo Biology." *Proceedings of the National Academy of Sciences of the United States of America 97*(9):4424–4425.

Gottlieb, A. 2000. "Luring Your Child into This Life: A Beng Path for Infant Care." In J. S. DeLoache and A. Gottlieb, eds., *A World of Babies: Imagined Childcare Guides for Seven Societies.* Cambridge and New York: Cambridge University Press.

Gould, S. J., and Lewontin, R. C. 1979. "The Spandrels of San Marco and the Panglossian Paradigm: A Critique of the Adaptationist Programme." Proceedings of the Royal Society of London. Series B. *Biological Science 205*(1161):581–598.

Grafton, S. T., Fadiga, L., Arbib, M. A., and Rizzolatti, G. 1997. "Premotor Cortex Activation During Observation and Naming of Familiar Tools." *Neuroimage 6*(4):231–236.

Greenfield, P. M., and Smith, J. H. 1976. *The Structure of Communication in Early Language Development.* New York: Academic Press.

Guthrie, R. D. 2006. *The Nature of Paleolithic Art.* Chicago: University of Chicago Press.

Halverson, H. 1937. "Studies of the Grasping Responses of Early Infancy: I." *Journal of Genetic Psychology 51*:371–392.

Hamlin, J. K., Wynn, K., and Bloom, P. 2007. "Social Evaluation by Preverbal Infants." *Nature 450*(7169):557–559.

Harlow, H. F. 1958. "The Nature of Love." *American Psychologist 13*:573–685.

Harnad, S. 1996. "The Origin of Words: A Psychophysical Hypothesis." In B. Velichkovsky and D. M. Rumbaugh, eds. *Communicating Meaning: The Evolution and Development of Language.* Hillsdale, NJ: Erlbaum.

Hauser, M. D. 1996. *The Evolution of Communication.* Cambridge, MA: MIT Press.

Hauser, M. D., Chomsky, N, and Fitch W. T. 2002. "The Faculty of Language: What Is It, Who Has It, and How Did It Evolve?" *Science 298*(5598):1569–1579.

Hawkes, K. 2003. "Grandmothers and the Evolution of Human Longevity." *American Journal of Human Biology 15*(3):380–400.

Heath, S. B. 1983. *Ways with Words: Language, Life, and Work in Communities and Classrooms.* Cambridge and New York: Cambridge University Press.

Herrmann, E., Call, J., Hernandez-Lloreda, M.V., Hare, B., and Tomasello, M. 2007. "Humans Have Evolved Specialized Skills of Social Cognition: The Cultural Intelligence Hypothesis." *Science 317*(5843):1360–1366.

Hewes, G. W. 1973. "Primate Communication and the Gestural Origin of Language." *Current Anthropology 14*:5–24.

Hirsh-Pasek, K., and Treiman, R. 1982. "Doggerel: Motherese in a New Context." *Journal of Child Language 9*(1):229–237.

Ho, Y. C., Cheung, M. C., and Chan, A. S. 2003. "Music Training Improves Verbal but Not Visual Memory: Cross-Sectional and Longitudinal Explorations in Children." *Neuropsychology 17*(3):439–450.

Hofman, M. A. 2001. "Brain Evolution in Hominids: Are We at the End of the Road?" In D. Falk and K. R. Gibson, eds., *Evolutionary Anatomy of the Primate Cerebral Cortex.* Cambridge, UK: Cambridge University Press.

Hohmann, G., and Fruth, B. 2008. "New Records on Prey Capture and Meat Eating by Bonobos at Lui Kotale, Salonga National Park, Democratic Republic of Congo." *Folia Primatologica* (Basel) *79*(2):103–110.

Holowka, S., and Petitto, L. A. 2002. "Left Hemisphere Cerebral Specialization for Babies While Babbling." *Science 297*(5586):1515.

Hopkins, M. 2007. "Babies Can Spot Nice and Nasty Characters." *Nature.* www.nature.com/news/2007/071121/full/news.2007.278.html

Horne, P. J., and Lowe, C. F. 1996. "On the Origins of Naming and Other Symbolic Behavior." *Journal of the Experimental Analysis of Behavior 65*(1):185–241.

Hrdy, S. B. 1994. "Fitness Tradeoffs in the History and Evoluton of Delegated Mothering with Special Reference to Wet-Nursing, Abandonment, and Infanticide." In S. Parmigiani and F. Vom Saal, eds., *Infanticide and Parental Care.* London: Harwood Academic.

Hrdy, S. B. 1999. *Mother Nature: Maternal Instincts and How They Shape the Human Species.* New York: Pantheon.

Huynh-Nhu, L. 2000. "Never Leave Your Little One Alone: Raising an Ifaluk Child." In J. S. DeLoache and A. Gottlieb, eds., *A World of Babies: Imagined Childcare Guides for Seven Societies.* Cambridge and New York: Cambridge University Press.

Iacoboni, M., Woods, R. P., Brass, M., Bekkering, H., Mazziotta, J. C., and Riz-
    zolatti, G. 1999. "Cortical Mechanisms of Human Imitation." *Science*
    286(5449):2526–2528.

Imada, T., Zhang, Y., Cheour, M., Taulu, S., Ahonen, A., and Kuhl, P. K. 2006.
    "Infant Speech Perception Activates Broca's area: A Developmental Mag-
    netoencephalography Study." *Neuroreport* *17*(10):957–962.

Iverson, J. M., Capirci, O., Longobardi, E., and Caselli, M. C. 1999. "Gesturing
    in Mother-Child Interactions." *Cognitive Development* *14*:57–75.

Iverson, J. M., and Goldin-Meadow, S. 2005. "Gesture Paves the Way for Lan-
    guage Development." *Psychological Science* *16*(5):367–371.

Jackendoff, R., and Pinker, S. 2005. "The Nature of the Language Faculty and
    Its Implications for Evolution of Language (Reply to Fitch, Hauser, and
    Chomsky)." *Cognition* *97*(2):211–225.

Johnson, M. C. 2000. "The View from the Wuro: A Guide to Child Rearing for
    Fulani Parents." In J. S. DeLoache and A. Gottlieb, eds., *A World of Babies:
    Imagined Childcare Guides for Seven Societies*. Cambridge and New York:
    Cambridge University Press.

Jones, S., Martin R. D., and Pilbeam, D. R. 1992. *The Cambridge Encyclopedia of
    Human Evolution*. Cambridge and New York: Cambridge University Press.

Jusczyk, P. W., Houston, D. M., and Newsome, M. 1999. "The Beginnings of
    Word Segmentation in English-Learning Infants." *Cognitive Psychology*
    39(3–4):159–207.

Kano, T. 1992. *The Last Ape: Pygmy Chimpanzee Behavior and Ecology*. Stanford,
    CA: Stanford University Press.

Karmiloff, K., and Karmiloff-Smith, A. 2001. *Pathways to Language: From Fetus
    to Adolescent*. Cambridge, MA: Harvard University Press.

Kawai, M. 1965. "Newly Acquired Pre-Cultural Behavior of the Natural Troop
    of Japanese Monkeys on Koshima Islet." *Primates* *6*(1):1–30.

Ke, J., Minett, J. W., Au, C.-P., and Wang, W. S.-Y. 2002. "Self-Organization and
    Selection in the Emergence of Vocabulary." *Complexity* *7*(3):41–54.

Kellogg, R., Knoll, M., and Kugler, J. 1965. "Form-Similarity Between
    Phosphenes of Adults and Pre-School Children's Scribblings." *Nature*
    208(5015):1129–1130.

Kempe, V., and Brooks, P. 2001. "The Role of Diminutives in the Acquisition of
    Russian Gender: Can Elements of Child-Directed Speech Aid in Learn-
    ing Morphology?" *Language Learning* *51*:221–256.

Kendon, A. 1988. "How Gestures Can Become Like Words." In F. Poyatos, ed.,
    Crosscultural Perspectives in Nonverbal Communication. Toronto: Hogrefe.

Keysers, C., Wicker, B., Gazzola, V., Anton, J. L., Fogassi, L., and Gallese, V. 2004.
    "A Touching Sight: SII/PV Activation During the Observation and Ex-
    perience of Touch." *Neuron* *42*(2):335–346.

Kirby, S. 1999. *Function, Selection and Innateness: The Emergence of Language Uni-
    versals*. New York: Oxford University Press.

Koda, H., Oyakawa, C., Kato, A., and Masataka, N. 2007. "Experimental Evi-
    dence for the Volitional Control of Vocal Production in an Immature
    Gibbon." *Behaviour* *144*(6):681–692.

Kojima, S. 2003. *A Search for the Origins of Human Speech: Auditory and Vocal Functions of the Chimpanzee.* Kyoto, Japan, and Portland, OR: Kyoto University Press.

Kuhl, P. K. 2000. "A New View of Language Acquisition." *Proceedings of the National Academy of Sciences of the United States of America* 97(22):11850–11857.

Kuhl, P. K. 2004. "Early Language Acquisition: Cracking the Speech Code." *Nature Reviews Neuroscience* 5(11):831–843.

Kuhl, P. K., Andruski, J. E., Chistovich, I. A., Chistovich, L. A., Kozhevnikova, E. V., Ryskina, V. L., Stolyarova, E. I., Sundberg, U., and Lacerda, F. 1997. "Cross-Language Analysis of Phonetic Units in Language Addressed to Infants." *Science* 277(5326):684–686.

Kuhl, P. K., Coffey-Corina, S., Padden, D., and Dawson, G. 2005. "Links Between Social and Linguistic Processing of Speech in Preschool Children with Autism: Behavioral and Electrophysiological Measures." *Developmental Science* 8(1):F1–F12.

Kuhl, P., and Meltzoff, A. N. 1988. "Speech as an Intermodal Object of Perception." In A. Yonas, ed. *Perceptual Development in Infancy: The Minnesota Symposium on Child Phonology.* Hillsdale, NJ: Erlbaum.

Ladefoged, P. 2004. *Vowels and Consonants.* Oxford, UK: Blackwell.

Ladygina-Kohts, N. N., de Waal, F., Vekker, B., and Yerkes Regional Primate Research Center. 2002. *Infant Chimpanzee and Human Child: A Classic 1935 Comparative Study of Ape Emotions and Intelligence.* Oxford; New York: Oxford University Press.

Lander, E. S., Linton, L. M., Birren, B., Nusbaum, C., Zody, M. C., Baldwin, J., Devon, K., Dewar, K., Doyle, M., FitzHugh, W., and others. 2001. "Initial Sequencing and Analysis of the Human Genome." *Nature* 409(6822):860–921.

Larson, S. G., Jungers, W. L., Morwood, M. J., Sutikna, T., Jatmiko, Saptomo E. W., Due, R. A., and Djubiantono, T. 2007. "*Homo floresiensis* and the Evolution of the Hominin Shoulder." *Journal of Human Evolution* 53(6):718–731.

Leighton, D. 1987. "Gibbons: Territoriality and Monogamy." In B. Smuts, D. L. Cheney, R. M. Seyfarth, R. W. Wrangham, and T. T. Struhsaker, eds., *Primate Societies.* Chicago: University of Chicago Press.

Levitt, A. G., and Utman, J. G. 1992. "From Babbling Towards the Sound Systems of English and French: A Longitudinal Two-Case Study." *Journal of Child Language* 19(1):19–49.

Liegeois, F., Baldeweg, T., Connelly, A., Gadian, D. G., Mishkin, M., and Vargha-Khadem, F. 2003. "Language fMRI Abnormalities Associated with FOXP2 Gene Mutation." *Nature Neuroscience* 6(11):1230–1237.

Linton, S. 1971. "Woman the Gatherer: Male Bias in Anthropology. In W. Jacobs, ed., *Women in Cross-Cultural Perspective.* Champaign-Urbana: University of Illinois Press.

Liu, H.-M., Kuhl, P. K., and Tsao, F.-M. 2003. "An Association Between Mothers' Speech Clarity and Infants' Speech Discrimination Skills." www3.interscience.wiley.com/journal/118852092/abstract?CRETRY=1&SRETRY=0

Longhi, E., and Karmiloff-Smith, A. 2004. "In the Beginning Was the Song: The Complex Multimodal Timing of Mother-Infant Musical Interaction." *Behavioral and Brain Sciences* 27:516–517.

Lorberbaum, J. P., Newman, J. D., Horwitz, A. R., Dubno, J. R., Lydiard, R. B., Hamner, M. B., Bohning, D. E., and George, M. S. 2002. "A Potential Role for Thalamocingulate Circuitry in Human Maternal Behavior." *Biological Psychiatry* 51(6):431–445.

Lordkipanidze, D., Jashashvili, T., Vekua, A., Ponce de Leon, M. S., Zollikofer, C. P., Rightmire, G. P., Pontzer, H., Ferring, R., Oms, O., Tappen, M., and others. 2007. "Postcranial Evidence from Early *Homo* from Dmanisi, Georgia." *Nature* 449(7160):305–310.

MacNeilage, P. F. 2000. "The Explanation of 'Mama.'" *Behavioral and Brain Sciences* 23:440–441.

MacNeilage, P. F. 1998. "The Frame/Content Theory of Evolution of Speech Production." *Behavioral and Brain Sciences* 21(4):499–511; discussion 511–446.

Maess, B., Koelsch, S., Gunter, T. C., and Friederici, A. D. 2001. "Musical Syntax Is Processed in Broca's Area: An MEG Study." *Nature Neuroscience* 4(5):540–545.

Maestripieri, D., and Call, J. 1996. "Mother-Infant Communication in Primates." *Advances in the Study of Behavior* 25:613–642.

Mania, D., and Mania, U. 1988. "Deliberate Engravings on Bone Artefacts of *Homo erectus*." *Rock Art Research* 5:91–107.

Marler, P. 1968. "Aggregation and Dispersal: Two Functions in Primate Communication." In P. C. Jay, ed., *Primates: Studies in Adaptation and Variability*. New York: Holt, Rhinehart.

Marlowe, F. W. 2005. "Hunter-Gatherers and Human Evolution." *Evolutionary Anthropology* 14(2):54–67.

Marlowe, F. W. 2006. "Central Place Provisioning: The Hadza as an Example." In G. Hohmann, M. M. Robbins, and C. Boesch, eds., *Feeding Ecology in Apes and Other Primates: Ecological, Physical and Behavioral Aspects*. Cambridge, UK: Cambridge University Press.

Masataka, N. 1998. "Perception of Motherese in Japanese Sign Language by 6-Month-Old Hearing Infants." *Developmental Psychology* 34(2):241–246.

Masataka, N. 2003. *The Onset of Language*. Cambridge, UK, and New York: Cambridge University Press.

Masataka, N. 2007. "Music, Evolution, and Language." *Developmental Science* 10(1):35–39.

Mekel-Bobrov, N., and Lahn, B. T. 2006. "What Makes Us Human: Revisiting an Age-Old Question in the Genomic Era." *Journal of Biomedical Discovery and Collaboration* 1:18.

Meltzoff, A. N. 1988. "Imitation, Objects, Tools, and the Rudiments of Language in Human Ontogeny." *Human Evolution* 3(1–2):45–64.

Meltzoff, A. N. 2007. "'Like Me': A Foundation for Social Cognition." *Developmental Science* 10(1):126–134.

Meltzoff, A. N., and Decety, J. 2003. "What Imitation Tells Us About Social Cognition: A Rapprochement Between Developmental Psychology and Cognitive Neuroscience." Philosophical Transactions of the Royal Society of Lond. Series B. *Biological Sciences 358*(1431):491–500.

Messinger, D. S., and Fogel, A. 1998. "Give and Take: The Development of Conventional Infant Gestures." *Merrill-Palmer Quarterly 44*(n4):566–590.

Miller, G. 2004. "Society for Neuroscience Meeting. Listen, Baby." *Science 306*(5699):1127.

Miller, G. F. 2000. "Evolution of Music Through Sexual Selection." In N. Wallin, B. Merker, and S. Brown, eds., *The Origins of Music*. Cambridge, MA: MIT Press.

Mills, M. T. 2004. *Phonological Features of African-American Vernacular English in Child-Directed Versus Adult-Directed Speech* (master's thesis). Columbus: Ohio State University.

Mithen, S. J. 2006. *The Singing Neanderthals: The Origins of Music, Language, Mind, and Body*. Cambridge, MA: Harvard University Press.

Monnot, M. 1999. "Function of Infant-Directed Speech." *Human Nature: An Interdisciplinary Biosocial Perspective 10:*415–443.

Morton, J. B., and Trehub, S. E. 2001. "Children's Understanding of Emotion in Speech." *Child Development 72*(3):834–843.

Morwood, M. J., Soejono, R. P., Roberts, R. G., Sutikna, T., Turney, C. S., Westaway, K. E., Rink, W. J., Zhao, J. X., van den Bergh, G. D., Due, R. A., and others. 2004. "Archaeology and Age of a New Hominin from Flores in Eastern Indonesia." *Nature 431*(7012):1087–1091.

Myowa-Yamakoshi, M, Tomonaga, M., Tanaka, M., and Matsuzawa, T. 2004. "Imitation in Neonatal Chimpanzees *(Pan troglodytes)*." *Developmental Science 7*(4):437–442.

Naigles, L. R., and Hoff-Ginsberg, E. 1998. "Why Are Some Verbs Learned Before Other Verbs? Effects of Input Frequency and Structure on Children's Early Verb Use." *Journal of Child Language 25*(1):95–120.

Nakamichi, M., Kato, E., Kojima, Y., and Itoigawa, N. 1998. "Carrying and Washing of Grass Roots by Free-Ranging Japanese Macaques at Katsuyama." *Folia Primatologica* (Basel) *69*(1):35–40.

Newman, J. D. 2004. "Motherese by Any Other Name: Mother-Infant Communication in Non-Hominin Mammals." *Behavioral and Brain Sciences 27:*519–520.

Nicolson, N A. 1977. "A Comparison of Early Behavioral Developments in Wild and Captive Chimpanzees." In S. Chevalier-Skolnikoff and F. E. Poirier, eds., *Primate Bio-social Development*. New York: Garland.

Nettl, B. 2000. "An Ethnomusicologist Contemplates Universals in Musical Sound and Musical Culture." In N. Wallin, B. Merker, and S. Brown, eds., *The Origins of Music*. Cambridge, MA: MIT Press.

Ochs, E. 1992. "Indexing Gender." In A. Duranti and C. Goodwin, eds., *Rethinking Context: Language as an Interactive Phenomenon*. Cambridge: Cambridge University Press.

Ochs, E. 1982. "Talking to Children in Western Samoa." *Language in Society* *11*:77–104.

Ochs, E., and Schieffelin, B. 1984. "Language Acquisition and Socialization: Three Developmental Stories. In R. Schweder and R. LeVine, eds., *Culture Theory: Mind, Self, and Emotion*. Cambridge: Cambridge University Press.

Panksepp, J., and Bernatzky, G. 2002. "Emotional Sounds and the Brain: The Neuro-Affective Foundations of Musical Appreciation." *Behavioural Processes* *60*(2):133–155.

Papaeliou, C. F., and Trevarthen, C. 2006. "Prelinguistic Pitch Patterns Expressing 'Communication' and 'Apprehension.'" *Journal of Child Language* *33*(1):163–178.

Papousek, M., Papousek, H., and Symmes, D. 1991. "The Meaning of Melodies in Motherese in Tone and Stress Languages." *Infant Behavior and Development* *14*:415–440.

Patel, A. D. 2003. "Language, Music, Syntax and the Brain." *Nature Neuroscience* *6*(7):674–681.

Patel, A. D, Iversen, J. R., and Rosenberg, J. C. 2006. "Comparing the Rhythm and Melody of Speech and Music: The Case of British English and French." *Journal of the Acoustical Society of America* *119*(5 Pt 1):3034–3047.

Petitto, L. A., Holowka, S., Sergio, L. E., and Ostry, D. 2001. "Language Rhythms in Baby Hand Movements." *Nature* *413*(6851):35–36.

Phillips-Silver, J. and Trainor, L. J. 2005. "Feeling the Beat: Movement Influences Infant Rhythm Perception." *Science* *308*(5727):1430.

Pickford, M. 2006. "Paleoenvironments, Paleoecology, Adaptations, and the Origins of Bipedalism in Hominidae." In H. Ishida, R. Tuttle, M. Pickford, N. Ogihara, and N. Nakatsukasa, eds., *Human Origins and Environmental Backgrounds*. New York: Springer.

Pierroutsakos, S. L. 2000. "Infants of the Dreaming: A Warlpiri Guide to Child Care." In J. S. DeLoache and A. Gottlieb, eds., *A World of Babies: Imagined Childcare Guides for Seven Societies*. Cambridge and New York: Cambridge University Press.

Pinker, S. 1997. *How the Mind Works*. New York: W. W. Norton.

Pinker, S., and Jackendoff, R. 2005. "The Faculty of Language: What's Special About It?" *Cognition* *95*(2):201–236.

Plooij, F. X. 1984. *The Behavioral Development of Free-Living Chimpanzee Babies and Infants*. Norwood, NJ: Ablex.

Pollick, A. S., and de Waal, F. B. 2007. "Ape Gestures and Language Evolution." *Proceedings of the National Academy of Sciences of the United States of America* *104*(19):8184–8189.

Poulin-Dubois, D., Graham, S., and Sippola, L. 1995. "Early Lexical Development: The Contribution of Parental Labelling and Infants' Categorization Abilities." *Journal of Child Language* *22*(2):325–343.

Preuss, T. 2001. "The Discovery of Cerebral Diversity: An Unwelcome Scientific Revolution." In D. Falk and K. Gibson, eds., *Evolutionary Anatomy of the Primate Cerebral Cortex*. Cambridge, UK: Cambridge University Press.

Preuss, T. 2006. "Who's Afraid of *Homo Sapiens?*" *Journal of Biomedical Discovery and Collaboration 1.* www.j-biomed-discovery.com/content/1/1/17

Preuss, T. 2007. "Evolutionary Specializations of Primate Brain Systems." In M. Ravosa and M. Dagasto, eds., *Primate Origins: Adaptations and Evolution.* New York: Springer.

Provine, R. R. 2000. *Laughter: A Scientific Investigation.* New York: Viking.

Provine, R. R. 2004. "Walkie-Talkie Evolution: Bipedalism and Vocal Production." *Behavioral and Brain Sciences* 27:520–521.

Pruetz, J. D., and Bertolani, P. 2007. "Savanna Chimpanzees, *Pan troglodytes verus*, Hunt with Tools." *Current Biology* 17(5):412–417.

Pusey, A., Williams, J., and Goodall, J. 1997. "The Influence of Dominance Rank on the Reproductive Success of Female Chimpanzees." *Science* 277(5327):828–831.

Pye, C. 1986. "Quiche Mayan Speech to Children." *Journal of Child Language* 13(1):85–100.

Pye, C. 1991. "The Acquisition of K'iché Maya." In I. Slobin, ed., *The Crosslinguistic Study of Language Acquisition. Vol 3.* Hillsdale, NJ: Erlbaum.

Ratner, N. B., and Pye C. 1984. "Higher Pitch in BT Is Not Universal: Acoustic Evidence from Quiche Mayan." *Journal of Child Language* 11(3):515–522.

Rauscher, F. H. 2002. "Mozart and the Mind: Factual and Fictional Effects of Musical Enrichment." In J. Aronson, ed., *Improving Academic Achievement: Impact of Psychological Factors on Education.* New York: Academic Press.

Rauscher, F. H., and Shaw, G. L. 1998. "Key Components of the Mozart Effect." *Perceptual and Motor Skills* 86(3 Pt 1):835–841.

Rizzolatti, G., and Arbib, M. A. 1998. "Language Within Our Grasp." *Trends in Neurosciences* 21(5):188–194.

Rizzolatti, G., and Craighero, L. 2004. "The Mirror-Neuron System." *Annual Review of Neuroscience* 27:169–192.

Rome-Flanders, T., and Cronk, C. 1995. "A Longitudinal Study of Infant Vocalizations During Mother-Infant Games." *Journal of Child Language* 22(2):259–274.

Rosenberg, K. R., and Trevathan, W. R. 2001. "The Evolution of Human Birth." *Scientific American* 285(5):72–77.

Ruff, C. B., Trinkaus, E., and Holliday, T. W. 1997. "Body Mass and Encephalization in Pleistocene *Homo.*" *Nature* 387(6629):173–176.

Sakai, K. L. 2005. "Language Acquisition and Brain Development." *Science* 310(5749):815–819.

Schaller, G. B. 1963. *The Mountain Gorilla; Ecology and Behavior.* Chicago: University of Chicago Press.

Schieffelin, B. B. 1990. *The Give and Take of Everyday Life: Language Socialization of Kaluli Children.* Cambridge and New York: Cambridge University Press.

Schmidt, K., and Cohn, J. F. 2001. "Human Facial Expressions as Adaptations: Evolutionary Questions in Facial Expression Research." *Yearbook of Physical Anthropology* 44:3–24.

Schultz, A. H. 1969. *The Life of Primates.* New York: Universe Books.

Semendeferi, K. 2001. "Advances in the Study of Hominid Brain Evolution: Magnetic Resonance Imaging (MRI) and 3-D Reconstruction." In D. Falk and K. Gibson, eds., *Evolutionary Anatomy of the Primate Cerebral Cortex*. Cambridge, UK: Cambridge University Press.

Senghas, A., Kita, S., and Ozyurek, A. 2004. "Children Creating Core Properties of Language: Evidence from an Emerging Sign Language in Nicaragua." *Science 305*(5691):1779–1782.

Sharpe, K., and Van Gelder, L. 2006. "Evidence for Cave Marking by Palaeolithic Children." *Antiquity 80*(310):937–947.

Sheridan, S. R. 2005. "A Theory of Marks and Mind: The Effect of Notational Systems on Hominid Brain Evolution and Child Development with an Emphasis on Exchanges Between Mothers and Children." *Medical Hypotheses 64*(2):417–427.

Sheridan, S. R. 2001. Very Young Children's Drawings and Human Consciousness: The Scribble Hypothesis, a Plea for Brain-Compatible Teaching and Learning (poster). Toward a Science of Consciousness Conference. Skovde, Sweden.

Shu, W., Cho, J. Y., Jiang, Y., Zhang, M., Weisz, D., Elder, G. A., Schmeidler, J., De Gasperi, R., Sosa, M. A., Rabidou, D., and others. 2005. "Altered Ultrasonic Vocalization in Mice with a Disruption in the Foxp2 Gene." *Proceedings of the National Academy of Sciences of the United States of America 102*(27):9643–9648.

Simpson, S. W., Quade, J., Levin, N. E., Butler, R., Dupont-Nivet, G., Everett, M., and Semaw, S. 2008. "A Female *Homo Erectus* Pelvis from Gona, Ethiopia." *Science 322*(5904):1089–1092.

Sloan, C. P. 2006, Nov. "The Origin of Childhood." *National Geographic Magazine,* pp. 148–159.

Small, M. F. 2001. *Kids: How Biology and Culture Shape the Way We Raise Our Children*. New York: Doubleday.

Small, M. F. 1998. *Our Babies, Ourselves: How Biology and Culture Shape the Way We Parent*. New York: Anchor Books.

Soltis, J. 2004. "The Signal Functions of Early Infant Crying." *Behavioral and Brain Sciences 27*(4):443–458.

Standley, J. M. 2002. "A Meta-Analysis of the Efficacy of Music Therapy for Premature Infants." *Journal of Pediatric Nursing 17*(2):107–113.

Stanford, C. B. 1995, Jan. "To Catch a Colobus." *Natural History Magazine:*48–55.

Stern, D. N., Spieker, S., Barnett, R. K., and MacKain, K. 1983. "The Prosody of Maternal Speech: Infant Age and Context-Related Changes." *Journal of Child Language 10*(1):1–15.

Surbeck, M., and Hohmann, G. 2008. "Primate Hunting by Bonobos at LuiKotale, Salonga National Park." *Current Biology 18*(19):R906–907.

Tanaka, M., Tomonaga, M., and Matsuzawa, T. 2003. "Finger Drawing by Infant Chimpanzees (*Pan troglodytes*)." *Animal Cognition 6*(4):245–251.

Tembrock, G. 1971. *Biokommunikation. Informationsübertragung im biologischen Bereich, Teil II*. Berlin: Akademie-Verlag.

Thiessen, E. D., and Saffran, J. R. 2007. "Learning to Learn: Acquisition of Stress-Based Strategies for Word Segmentation." *Language Learning and Development* 3:75–102.

Thomas, G. V., and Silk, A. M. J. 1990. *An Introduction to the Psychology of Children's Drawings.* New York: New York University Press.

Tincoff, R., and Jusczyk, P. W. 1999. "Some Beginnings of Word Comprehension in Six-Month-Olds." *Psychological Science* 10:172–175.

Tomasello, M., and Camaioni, L. 1997. "A Comparison of the Gestural Communication of Apes and Human Infants." *Human Development* 40(1):7–24.

Tomasello, M., and Carpenter, M. 2007. "Shared Intentionality." *Developmental Science* 10(1):121–125.

Trainor, L. J., Austin, C. M., and Desjardins, R. N. 2000. "Is Infant-Directed Speech Prosody a Result of the Vocal Expression of Emotion?" *Psychological Science* 11(3):188–195.

Trainor, L. J., Clark, E. D., Huntley, A., and Adams, B. A. 1997. "The Acoustic Basis of Preferences for Infant-Directed Singing." *Infant Behavior and Development* 20(3):383–396.

Trehub, S. E. 2000. "Human Processing Predispositions and Musical Universals." In N. Wallin, B. Merker, and S. Brown, eds., *The Origins of Music.* Cambridge, MA: MIT Press.

Trehub, S. E. 2002, Sept. *Mothers Are Musical Mentors. Zero to Three* 23(1):19–22.

Trehub, S. E. 2003. "The Developmental Origins of Musicality." *Nature Neuroscience* 6(7):669–673.

Trehub, S. E., Trainor, L. J., and Unyk, A. M. 1993. "Music and Speech Processing in the First Year of Life." *Advances in Child Development and Behavior* 24:1–35.

Trehub, S. E., Unyk, A. M., Kamenetsky, S. B., Hill, D. S., Trainor, L. J., Henderson, J. L., and Saraza, M. 1997. "Mothers' and Fathers' Singing to Infants." *Developmental Psychology* 33(3):500–507.

Trehub, S. E., Unyk, A. M., and Trainor, L. J. 1993. "Adults Identify Infant-Directed Music Across Cultures." *Infant Behavior and Development* 16:193–211.

Tsao, F. M., Liu, H. M., and Kuhl, P. K. 2004. "Speech Perception in Infancy Predicts Language Development in the Second Year of Life: A Longitudinal Study." *Child Development* 75(4):1067–1084.

Vanhaereny, M., d'Errico, F, Stringer, C., James, S. L., Todd, J. A., and Mienis, H. K. 2006. "Middle Paleolithic Shell Beads in Israel and Algeria." *Science* 312(5781):1785–1788.

Volkova, A., Trehub, S. E., and Schellenberg, E. G. 2006. "Infants' Memory for Musical Performances." *Developmental Science* 9(6):583–589.

Walker, A., and Ruff, C. B. 1993. The Reconstruction of the Pelvis. In A. Walker and R. Leakey, eds., *The Nariokotome Homo erectus Skeleton.* Cambridge, MA: Harvard University Press.

Wallace, I. J., Demes, B., Jungers, W. L., Alvero, M., and Su, A. 2008. *The Bipedalism of the Dmanisi Hominins: Pigeon-Toed Early Homo? American Journal of Physical Anthropology* 136(4):375–378.

Watson, J. 1928. *Psychological Care of Infant and Child.* New York: W. W. Norton.

Werker, J., and Tees, R. C. 1984. "Cross-Language Speech Perception: Evidence for Perceptual Reorganization During the First Year of Life." *Infant Behavior and Development* 7:49–63.

Wermke, K., Leising, D., and Stellzig-Eisenhauer, A. 2007. "Relation of Melody Complexity in Infants' Cries to Language Outcome in the Second Year of Life: A Longitudinal Study." *Clinical Linguistics and Phonetics* 21(11–12):961–973.

Wermke, K., and Mende, W. 2006. "Melody as a Primordial Legacy from Early Roots of Language." *Behavioral and Brain Sciences* 29:300.

Wermke, K., Mende, W., Manfredi, C., and Bruscaglioni, P. 2002. "Developmental Aspects of Infant's Cry Melody and Formants." *Medical Engineering and Physics* 24(7–8):501–514.

Wheeler, P. 1988, May 12. "Stand Tall and Stay Cool." *New Scientist:* 62–65.

Whiten, A., and Byrne, R. W. 1997. *Machiavellian Intelligence II: Extensions and Evaluations.* Cambridge/New York: Cambridge University Press.

Wilcox, S. 2004. "Language from Gesture." *Behavioral and Brain Sciences* 27:525–526.

Wilson, S. M., Saygin, A. P., Sereno, M. I., and Iacoboni, M. 2004. "Listening to Speech Activates Motor Areas Involved in Speech Production." *Nature Neuroscience* 7(7):701–702.

Wolff, P. 1969. "The Natural History of Crying and Other Vocalizations in Early Infancy." In B. Ross, ed., *Determinants of Infant Behavior.* London: Methuen.

Wright, C. M., and Parkinson, K. N. 2004. "Postnatal Weight Loss in Term Infants: What Is Normal and Do Growth Charts Allow for It?" *Archives of Disease in Childhood Fetal and Neonatal Edition* 89(3):F254–257.

Yamagata, K. 2001. "Emergence of Representational Activity During the Early Drawing Stage: Process Analysis." *Japanese Psychological Research* 43:130–140.

Yu, X. J., Zheng, H. K., Wang, J., Wang, W., and Su, B. 2006. "Detecting Lineage-Specific Adaptive Evolution of Brain-Expressed Genes in Human Using Rhesus Macaque as Outgroup." *Genomics* 88(6):745–751.

Zatorre, R. J., Perry, D. W., Beckett, C. A., Westbury, C. F., and Evans, A. C. 1998. "Functional Anatomy of Musical Processing in Listeners with Absolute Pitch and Relative Pitch." *Proceedings of the National Academy of Sciences of the United States of America* 95(6):3172–3177.

Zihlman, A. 1985. "Gathering Stories for Hunting Human Nature." *Feminist Studies* 11:365–377.

INDEX